# SpringerBriefs in Business



For further volumes:
http://www.springer.com/series/8860

Marco Guerzoni

# Product Variety in Automotive Industry

## Understanding Niche Markets in America

Springer

Marco Guerzoni
Department of Economics and Statistics
University of Turin
Turin, Torino
Italy

ISSN 2191-5482 ISSN 2191-5490 (electronic)
ISBN 978-3-319-01906-2 ISBN 978-3-319-01907-9 (eBook)
DOI 10.1007/978-3-319-01907-9
Springer Cham Heidelberg New York Dordrecht London

Library of Congress Control Number: 2013947350



Printed on acid-free paper

Springer is part of Springer Science+Business Media (www.springer.com)

# Summary

The subject of this book is product variety in the United States automobile industry. Specifically, it aims to understand the organization and the limits of the generation of variety at the industry level.

This sector shows a dual structure: on the one hand, there are relatively few and large companies producing cars that apparently achieve a degree of market power through the differentiation of their production. On the other hand, a relatively small niche market with distinct and smaller producers offers specialty equipment to enhance the performance, appearance, and handling of vehicles. This peculiarity allows an original type of empirical analysis of variety generation defined niche market analysis. By comparing the variety produced by the large companies with that provided by the niche market, it will be possible to understand more deeply the nature and operation of niche markets and to provide fresh empirical evidence on the limits of variety generation at the industry. Results have implications for economic theory and management and call for further research on this topic.

# Contents

# Introduction

This book is about product variety in the U.S. automobile industry. Specifically, it aims to understand the organization and the limits of the generation of variety at the industry level.

This topic is worth addressing for two main reasons. The first is linked to the specific characteristics of the sector of the automobile industry, which shows a dual structure in the production of variety: on the one hand there are relatively few and large companies producing cars that apparently achieve a degree of market power through the differentiation of their production. On the other hand, a relatively small niche market with distinct and smaller producers offers specialty equipment to enhance the performance, appearance and handling of vehicles. **By analyzing the nature of this niche market I aim to further develop the theory of niche market and to provide fresh empirical evidence on the nature and operation of these markets.** There already have been attempts to address the nature of niche markets and their links with the generation of variety empirically. Scholars of industrial ecology have tried to extend their analysis to the evolution of industrial organization in specific industries such as newspapers and wine producers (Boone et al. 2002; Swaminathan 2001, both based on the seminal contribution by Carroll 1985). Works with an evolutionary flavor have focused on bifurcation of trajectories and on the emerging of new technology by looking at market's niches (e.g., Frenken et al. 1999 analyzed aircrafts, helicopters, motorcycles, and microcomputers).

However the case for the automobile sector is rather different. This is because the specialty equipment niche market does not develop a particular version of a product as it is the case presented in the ecological literature. In addition, the specialty equipment market is not a different market *per se* and neither it represents a different technological trajectory. Its nature is to provide consumers with tools to customise their product. This peculiarity allows an original type of analysis of variety generation. As explained in detail in Chap. 3, by comparing the variety produced by the large companies with that provided by the niche market, it will be possible to understand more deeply the nature of variety.

Thus, **a second aim of this book is to shed empirical light on the limits of variety's generation.** A body of theory (Henderson and Clark 1990) studies improvements in the product design and describes limits in variety's generation as intrinsically related with the hierarchical structure of the product's design.

A different stream of literature (Lancaster 1966, 1971), that considers variety mainly as product differentiation, focuses on the comparison between the cost structure in producing variety and in the incentives arising from the value given to heterogeneity by the demand side of the market. Here the limits to variety are exogenous to the product, whose characteristics are flexible and, theoretically, do not have upper boundaries in the various forms they can assume. **This book will empirically distinguish between these explanations concluding that one emerges as the most convincing for the US automobile industry and examining to what extent the two approaches can coexist.**

The second reason to address this issue is not industry specific: the analysis of variety, especially linked to the existence of niche markets is crucial for the understanding of economic development.

The most convincing explanation for the existence of the fringe of small firms is the need to supply some segments of demand representing preferences that are too diverse to accept the standard product. It is reasonable to assume that consumers with these preferences are not simply odd people with weird tastes; on the contrary, their diversity is often linked with a higher degree of sophistication. There are different types of these users, which, as highlighted in the book, might be the real engine of qualitative innovation.

This qualitative innovation and, therefore, variety generation are crucial for the economic growth. The evolution of mass production in the last century has been a continuous attempt to increase technology's productivity through process innovation (Piore and Sabel 1984; Abernathy and Utterback 1978). If productivity steadily increases and demand reaches a saturation point, the economy can incur in an imbalance because it produces the amount of good and service required with decreasing amounts of inputs, including labour. The growth in variety counterbalances this bottleneck by creating new sectors that employ new inputs. Since variety generation is crucial for economic development and since niche markets seem to play an important role, it is worth shedding some light on the phenomenon: **if any result of this research can be generalized, it can have an importance in the understanding of the process of growth and welfare creation, the ultimate questions of economic analysis.**

The first chapter is about the evolution of variety in the automobile industry and describes the context in which the analysis is carried out. The second chapter will address the theoretical basis for an economic analysis of variety and provide guidance for the data analysis. Moreover it will develop few concepts that will be used in the empirical part. The third chapter is divided into two parts. It will first describe the new methodology applied in book for the data analysis and, thereafter, will proceed with the empirical analysis of variety generation. Finally, conclusions will resume the aims and the achievements of the book.

## References

Abernathy, W. J. & Utterback, J. M. (1975). Patterns of industrial innovation. *Technology Review 80*(7), 40–47.

Boone, C., Carroll, G. R., & van Witteloostuijn, A. (2002). Environmental resource distributions and the market partitioning of Dutch daily newspaper organizations. *American Sociological Review, 67*, 408–431.

Carroll, G. R. (1985). Concentration and specialization: Dynamics of niche width in populations of organizations. *American Journal of Sociology*, *90*, 1262–1283.

Frenken, K., Saviotti, P. P., & Trommetter, M. (1999). Variety and niche creation in aircraft, helicopters, motorcycles and microcomputers. *Research Policy*, *28*(5), 469–488.

Henderson, R., & Clark, K. (1990). Architectural innovation. *Administrative Science Quarterly*, *35*, 9–30.

Lancaster, K. J. (1966). A new approach to consumer theory. *Journal of Political Economy*, *4*, 132–157.

Lancaster, K. J. (1971). *Consumer demand: A new approach*. New York: Columbia University Press.

Piore, J. M., & Sabel, C. F. (1984). *The second industrial divide. Possibilities for prosperity*. New York: Basic Books.

Swaminathan, A. (2001). Resource partitioning and the evolution of specialist organization: The role of location and identity in the U.S. wine industry. *Academy of Management Journal*, *44*, 1169–1185.

# Chapter 1
# Variety in the Automobile Industry

The automobile manufacturing industry, along with its dealers and suppliers, is one of the industry that most contributes to the American economic growth: it generates about 3.7 % of American GDP by producing and selling new light vehicles, a level of output higher than any other industry of comparable aggregation. In 1997, the industry invested 18.4 billion dollars in research and developed and exported 74 billion dollars (source: Alliance of Automobile Manufacturer website).

In the automobile industry the big three (Ford Motor Company, General Motor Corporation, Chrysler) dominate the market. Ford Motor Company (Ford) and General Motor (GM) are the two world's largest car producers with a joint market share between 21 and 25 %. The big three together with a few European and Japanese manufacturers are responsible for the production of the 216,183,000 vehicles in use in US in 2001 (Fine et al. 1996). The figures are impressive not only in term on quantity produced; also the variety generated in the market is noticeable: in 2003, 34 companies in the American market produced hundreds of models of cars for 5 different market segments (passenger cars, pick-up trucks, mini vans and vans, sport utility vehicles, sports cars). For the passenger vehicle market alone, 148 different models were on offer. Moreover each model is available in several trims and each trim in different colours with an increasing possibility of customising the product with dozens of ad hoc additional features (Source: National Automobile Dealer Association website).

In the following paragraphs, by referring to the history of this sector, I discuss the factors that led to such a considerable variety generation.

## 1.1 Variety in the History of the Horseless Carriage

In the late 1890s the automobile ceased to be only a curiosity for mechanical engineers and became a product with a market steadily and rapidly growing. The industry in those early days was highly competitive because car producers were essentially assembler of parts and, therefore, only mechanical competencies and



M. Guerzoni, *Product Variety in Automotive Industry*,
SpringerBriefs in Business, DOI: 10.1007/978-3-319-01907-9_1,

little investment were required to enter the business. In 1900, according to the Census Bureau, 4,191 different models were produced and sold mainly by several small companies and shops. Among these small producers, there was an engineer with a firm trust in the future of the car as a mass market product: Henry Ford aimed to produce a car for the multitude, that was cheap enough to be bought by workers and solid enough to be driven on the country roads: *I want to build a car for the great multitude (…). But it will be so low in price that no man making a good salary will be unable to own one* (Henry Ford quoted in Wik 1972, p. 233). In 1908, when already 400,000 cars were registered in the United States, he introduced his Model T and set a landmark in the history of the industry. It was Ford's twentieth attempt, having started 5 years before. Ford's prediction was correct and within 10 years he was able to sell more than half a million cars in a single year (Rae 1965, p. 62). Because of its characteristics the Model T diffuses everywhere. Moreover, it was so versatile that someone believed it *could do everything except talk and climb a tree* (Wik 1972, p. 33). It was even used in agriculture: its 20-horsepower engine, just by attaching pulley to the crankshaft, could be used for instance for grinding grain, churning butter, shelling corn and by attaching some extra parts could be even converted into a snowmobile (Wik 1972, p. 33).

In order to minimize the cost of production and, therefore, to be able to sell his car for a low price (850$ in 1908) Henry Ford introduced a revolution in the manufacturing industry and marked the shift from craft production to mass production. The economic development can be conceived as a *centuries-old collective experiment in productive organization* (Piore and Sabel 1984, p. 4). For many industries, the end of nineteenth century observed a sharp industrial divide where two different forms of productive organizations were competing: craft production and mass production. Craft production originates in the idea that technology, by being an extension of worker's skill, should augment his ability to create variety. On the contrary, the foundation of mass production is the substitution of labour with capital. For this reason, it most often adopts the tendency of dividing labour into simple steps that could, in principle or eventually if not at present, be performed by dedicated machinery (Piore and Sabel 1984, p. 19). The human element is required only to the extent that it is necessary to control the machine. In the words of Marx *in handcraft […] the workman makes use of a tool, in the factory, the machine makes use of him* (Marx 1867; quoted in Piore and Sabel 1984, p. 310).

Ford understood that the key requirement to allow the shift to mechanization and, therefore, to mass production was

> (…) to make one automobile like another automobile. To make them all alike, to make them come from the factory just alike- just like one pin is like another pin when it comes from a pin factory (Henry Ford, quoted in Rae 1965, p. 61).

This allowed a sharp division of labour and the exploitation of economies of scale: in 1908, when he managed to have in his model perfect interchangeability among components, decided that the assembler would focus on a single task moving around the shop floor from vehicle to vehicle. One year later, again to maximize the exploitation of scale economies, Henry Ford decided to concentrate exclusively on the production of Model T and give up all other models. In 1913 he introduced the moving assembly line and he achieved a further improvement in productivity and reduction in price (Rae 1965). Only firms able to compete at this level of efficiency could survive. Before the production of Ford's Model T, the industry observed many firms entering the market. The peak of the entry was in 1907 with 82 firms entering the market. Thereafter a steady shake out took place, slow in the first 2 years and increasingly sharper afterwards. In 1910 the number of entries fell behind the number of exits. From 1909 to 1941 the number of firms dropped from 272 to 9 (Klepper 1996; Klepper 2001; Rae 1984).

The variety in industry decreased dramatically: not only few firms were left on the market, but in order to exploit economies of scale and stay in the market they were focusing on producing few different models. Moreover the key change in mass production is … *the complete and consistent interchangeability of parts and the simplicity of assembling to each other* (Womack et al. 1990, p. 12; also quoted in Barnes 2001). For this reason, each single unit produced was exactly alike, as Ford had forecasted. However, the car experienced noticeable improvements in terms of performance, security and comfort: the Model A in 1928 had twice the horsepower of Model T, the lower centre of gravity reduced the risk of turning over, and some models were offering the first optional items like radios (Rae 1965, p. 107).

It is possible, however, that the idea of mass production in the automobile industry has been largely misunderstood. Instead of the idea that mass production is about producing a large amount of low quality goods that are acceptable only because of their low price it may be that:

> the purpose of mass production is to combine increased output with decreased unit cost. It achieves this purpose by using manufacturing methods that have an exactness of coordination and synchronization and standard of accuracy and inter-changeability far beyond the capacity of the most skilled craftsman. The alleged sacrifice of quality to quantity is a myth (Rae 1965, p. 53).

This quotation, if taken as general rule for industry evolution, can be falsified by many counterexamples. In particular, in decorative art products will surely have a higher quality if craft produced by *Fine Art Workmen in Painting, Carving, Furniture and the Metals* as Morris and his pre-Raphaelite friends defined their activity (Morris 1934; Faulkner 1980). Nevertheless it is hardly contestable in the early days of the automobile industry: if there was any negative correlation between quality and quantity, the Model T, with 1,532,425 units of total sales in the first 8 years of production should have been the worst car ever produced.

However, as we have seen, if the sacrifice of quality to quantity was a myth, the introduction of mass consumption shrank inexorably the variety that was available on the market. At least in the beginning of the automobile industry, therefore, the real trade off was not between quantity and quantity but between variety and cost: *the consumer gets lower costs but at the expense of variety* (Womack et al. 1990, p. 13). In 1920, 50 % of vehicles in the world were a Model T. In the history of the "horseless carriage", the range of variety has never been that narrow again. From this standpoint, the history of the automobile industry can be seen as a successful attempt of increasing variety both in term of market strategy and process' technologies.

### 1.1.1 Generation of Variety and Market's Strategy

The company to implement first differentiation strategies was GM. Since the beginning of its history GM focused on market segmentation and product differentiation. This strategy was in the DNA of the company: if Ford emerged since the beginning as a very integrated company, GM was founded by William C. (Billy) Durant by merging together different producers: Buick Motor Company, Maxwell-Briscoe Cadillac, Oldsmobile, and Oakland. The result in marketing strategy of this multinomial nature has been a constant attempt to produce different models from the low price Chevrolet to luxury cars.

The differentiation strategy was also dictated by other two factors: first the absolute advantage of Ford in the market segment of Model T; secondly, the belief that demand for low cost car, the market for the Model T, would have been fulfilled increasingly by used cars. It is worth noting that GM first made efforts to integrated design and production with the purpose of standardizing the part used to assemble different types of vehicles. GM' strategy was successful and soon copied by the other mass producers (Piore and Sabel 1984, pp. 69–70). After dismissing the Model T, Ford broadened the range of products and Chrysler took a similar decision. Already in the 1950s, after the troublesome period of depression and the war, consumers had a vast choice of variety: different colours both for the exterior body and for the interior, different type of engines and hubcaps; radio, heater, and air-conditioner were optional items already available (Rae 1965). Thereafter the variety of cars' design and range of additional features has been rising steadily: several fashions emerged in the automobile industry, from the European compact in the 1960s until the S.U.V. in this decade. In order to achieve the level of variety existing nowadays two process and organizational innovations have been crucial. The first one started in Japan in the late 1950s; the second emerged slowly in the 1980s and 1990s. They are the rise of lean production as manufacturing best practice and the concept of modular architecture in the product design.

### 1.1.2 Generation of Variety and Production: Lean Production and Modularity

After the shift to mass production, a second important revolution in the organisation of production was the implementation in Japanese companies of lean production. In the 1950s, Japanese companies, in particular Toyota, considered entering the world automobile market, although there were aware that they could not compete against the large American and European mass producers for several reasons: in Japan the financial resources to buy the expensive machineries required by mass production were scarce and there was a ban on foreign direct investments in the motor industry; the labour was not cheap and there was no industrial reserve army of immigrants to employ at low wages; finally the Japanese market required much more variety than American one, ranged from luxury cars for public officers to small cars for the overpopulated Japanese cities (Womack et al. 1990; Barnes 2001). The solution was found by Toyota Motor Company and originally known as the Toyota System of Production (TSP), which became famous worldwide as lean production after the publication of the book "The machine that changed the world" by Womack et al. (1990) based on the five million dollars and 5 years M.I.T. research program on the automobile industry.

The genius of Eiji Toyoda and of his chief production engineer, Taiiche Ohno, made possible the development of lean production in Japan. In 1937, the Toyoda family founded the Toyota Motor Company and, in its first years of life, it did not succeed in successfully entering the cars' market. In 1950, after 13 years of existence, Toyota Motor Company had produced only 2,685 automobiles, whilst the Ford's Rouge plant in Detroit was producing 7,000 vehicles every day. Eiji and Ohno visited for 3 months the River Rouge plant. They acknowledge that it was not possible to utilise mass production techniques in Japan but that *there were some possibilities to improve the production system* (Eiji Toyoda, quoted in Womack et al. 1990, p. 37). They understood that one weakness of the system was the excessive presence of *Muda*, the Japanese world for waste. lean production is an attempt to increase the productivity by minimising waste (Cusumano 1988).

Lean production is based on three principles:

*Different organisation of labour*. The different organization of labour reflects the institutional conditions in the post-war Japanese economy. The new labour legislation guaranteed life employment to the work force. From one side, companies had to considered employees as a long-term investment and not as a variable cost. Employees, on the other hand, start considering themselves as active part of the company, willing to suggest solutions and not only to respond mechanically to problems and to obey to orders (Womack et al. 1990). A concrete example was the process of error detection and resolution. An error in mass production was considered as a random event. The worker, even in the presence of mistakes was not allowed to stop the assembly line to fix it, but at the end of the production process dedicated specialists investigated for errors and repaired them.

In lean production the system is working differently: workers are organised in teams that should performs different assembly activities. If the team identify an error, they are allowed to stop the assembly line both to fix the problem and to find the reason why it has occurred. In other words the mistake is considered as systematic and there is an attempt to explain its causes made directly by the workers. As soon Ohno introduced this new organization of production, the line was stopping all of the time. As time went by, however, workers gained experience in identifying the ultimate causes of a problem and the number of errors began to drop dramatically (Womack et al. p. 57).

*Different assembly process.* One of the biggest improvements in flexibility was due to new techniques of stamping metal parts from sheet steel. In mass production the stamping process took place on massive press lines where some heavy dies were set to shape the iron sheets. Since the dies in the press, which shaped the metal, weighted many tons, changing dies in an assembly line required time and specialist labour. The typical solution to minimize the problematic dies' changing was to dedicate each assembly line to the stamping of a single car's part. This techniques did not suit Japanese companies because of the high minimal scale required and because assembly lines should have been dedicated to the production of a single part. Conditions in Japan dictated a smaller scale of production and, relatedly, that only few press lines stamped all the parts needed.

The solution founded by Ohno has been to develop dies that were easy to change by making smaller batches and, then, to change them frequently. This allowed him to use few assembly lines to produce all the parts required. This solution worked efficiently and, as a by-product, it led to additional savings. First, the process eliminated the need for large inventories; secondly, making few parts before assembling them made the identification of errors and correction in later stampings easier.

*Different vertical industrial organization.* Finally, Ohno introduced major changes also in the relationship with suppliers and dealers for the final market. The extent of formal integration in western producers varied across firms and over time ranging from 25 % in small specialist firms like Porsche to about 100 % of Ford's Rouge plant before the Second World War. In Japan, Toyota did not perceive the problem of vertical integration as a "make or buy" decision. Toyota rather focused on the way they could guarantee a coordination among suppliers and between suppliers and assembler to improve quality and reduce price. Toyota managed to do it by holding the suppliers in long-term formal relationships, by acting as bank for loan and investment in new process, by decentralizing the design of components and by sharing personnel. On the marketing side, Toyota, aware of the flexibility and reliability the system gained with lean production, decided to employ a new philosophy in dealing with consumer by introducing the "aggressive selling".

The aim was to have complete integration between distribution and production in order to produce within a reasonable time (2 or 3 weeks) cars already sold by the dealers. This organisation became famous with the expression "Just in Time". The result was not only a further reduction of expensive inventories but also the possibility of producing a car for a specific customer with the requirements she demanded.

American companies did not immediately understand the potential for the variety generation offered by lean production because their Japanese competitors where concentrating on one or two models for each export market in order to minimize distribution costs: *They are making identical cans; if I did that I could have high quality and low cost, too* (Manager of a large cars' producer, quoted in Womack et al. 1990, p. 65). On the contrary, the lean production by reducing fixed cost allow Japanese producers to shorten the average product life and enter the market with more variety. The flexibility of Japanese plants made possible the production of two or three models per plant over a period of 3 or 4 years, while the Western competitors were, on average, producing one single model per plant over a period of 10 years. The Just in Time philosophy allowed them to produce cars more customised to users need. All these factors together increased the variety and allowed Japanese producers to meet better demand's requirements. Only in the late 1980s did it become clear that lean production was a superior way of manufacturing and American producers had to find their own way to achieve the equivalent flexibility through for example "Flexible Specialisation" (Piore and Sable 1984), "Post-Fordism" (Lipietz 1987), "New Competition" (Best 1990). Among the solutions, which emerged to be most flexible, modularity in product architecture's design was a new concept that most impinged on the generation of variety.

Modularity can be approach from three different perspectives: modularity in design, modularity in production and in organization (Calcagno 2002; Baldwin and Clark 1997, 2000; Salerno and Dias 2002).[1]

Alexander (1964) and Simon (1962) first introduced the concept of *modularity in design* as system configuration that improves the architecture of a complex system. Alexander considered design to be the fit between a context and a form, *id est* between a problem and its solutions. The presence of modules allows the designer to make her design more tailored to different and changing contexts. This flexibility, for instance, can be used to adapt the product to different market segments and increase the variety.

*Modularity in production* is a more practical concept and it depicts the idea of producing general interchangeable parts or modules that can be added during the assembly process to a product specific platform (Calcagno 2002). It is reasonable to assume that a product design is not integral or modular but that it varies in continuum between the two extremes. On the one hand, therefore, there are modules and on the other, a core of stable components conceived as integral. This core is the platform. Modules are added to the platform, through standard interfaces (Gawer and Cusumano 2002). Modularity in production, according to Starr (1965) can broaden the range of variety: inputs are used to produce different parts that are thereafter assembled together in the final product. His basic idea is to develop parts, which can be combined in the maximum number of ways (Starr 1965, p. 138). *Modularity in organization* involved the industrial structure at meso-level and however is not discussed further here since it is not crucial to explain the generation of variety in the automobile industry.

[1] This division follows Calcagno (2002) and Salerno and Dias (2002), however the distinction between modularity in design and modularity in production is common in the literature (Baldwin and Clark 1997 for a discussion on the categorisation).

For the automobile industry modularisation in design offers a way to reduce complexity (Camuffo 2002).

> Modularization means that, in the future, vehicles will probably result from the integration of a series of self-contained functional units with standardized interfaces within one or more standardized product architectures, units conceived, manufactured or supplied, and assembled as autonomous modules (Camuffo 2002, p. 3; referring to Helper et al. 1999).

For the moment, however, the automobile industry is still far from designing a modularised product: the product architecture is not modular but integral, there are no cross firm standardizations and only few cross product standardizations (Ulrich 1995; Camuffo 2002). On the contrary, modularity in production has been a reality. The new assembly plants of GM and Ford in Rio Grande do Sul, and the new Chrysler factory In Curitiba, Paranà, are concrete example of this tendency. There are also examples of cross firm modularity: the result of GM-FIAT joint venture was the production of a modular car under-body that it has been used for different brands in different market segment (Camuffo and Volpato 2001); Faurecia and Sommer Allibert Industrie invested in designing passenger compartment and door modules without appealing to specific brands (Murphy 1999; Farhi 1998; quoted in Camuffo 2002). Denso International America is producing integrated air–fuel adapt for all car makers (Camuffo 2002).

It could be argued that this kind of modularity in the automobile industry has been pull by new requirements in supply chain management after the rise of lean production and flexible production. The most important contributions along this line are Sako and Warburton (1999), Fujimoto and Takeishi (2001) and the review by Calcagno (2002). They address the issue both from the point of view of the assembler and from the supplier: for the assembler modularisation is a way to increase efficiency in the supply chain management; on the other hand, for the supplier modularity is essentially an opportunity to escape the monopsonistic situation, where she depends completely on the assembler's market power.

But modularity in production, also offers a great opportunity of producing variety (Starr 1965). Starr explained in a very influential article in the Harvard Business Review that when a product is mature, consumers tend to discard the *keeping up with the Joneses* (Dichter 1965), such that possession should emphasizes their distinction from other people. Modularity is the way, according to Starr, to provide the market with the diversity required. In Starr's seminal idea is rooted the concept of mass customisation, as developed at the end of 1980s by Davis (1987) and Pine (1993). They observed in the automobile industry, as in many others sectors, this tendency of delivering more and more variety and that modularity in production was one of the key driving forces of this process. They explained that

> no longer do the companies focus on producing standardized products or services for homogeneous markets: They begin reaching down closer to the individual. Greater and greater variety blends into more and more customised products (Chan 2001, p. 1; referring to Pine 1993).

In sum, from Ford's quotation *The Customer Can Have Any Color He Wants So Long As It's Black* to Build to Order,[2] the American automobile industry managed to overcome the initial trade-off between variety and costs.

## 1.2 The Persistent Dualism in Variety Generation: The Niche Market for Special Equipment

Nowadays, in the automobile industry, diversity is considerable both in term of market segmentation and personal customisation. However a paradox seems to emerge. *Reaching down closer to individuals* should imply the end of a niche market that provides a changing variety of oddments and opportunities to *do it yourself* (Pine 1993). However, the large companies are not absorbing the role provided by niche markets for generating variety. On the contrary, the American automobile industry is showing an increasing dualism in product variety among mass market and the niche market for special equipment.

Averitt (1968) first used the term dual economy to describe a dichotomy in the industry structure: on the one hand there is a core of large oligopolistic companies, on the other hand, the fringe of every industry has a plethora of many small firms. The former is the core of the economy, the firms are market maker because they not only can set price and quantity but they have also political power to influence the institutional setting. The latter, defined by Averitt as periphery, is close to the prototype of perfect competition in the orthodox theory: small firms, price takers, with no power to influence to market or the institutional setting. They were often taken as examples by orthodox theorists (Solow 1967) as counterexample to the Galbraith' (1967) and Marris' (1964) approach that strongly attacked the idea of perfect competition (Berger and Piore 1980). Recently, the availability of large longitudinal data showed that the uneven distribution of firm characteristics (size, profit, investment, market power, etc.) is a common pattern across industries and states (for a review of some stylised facts see Dosi et al. 1995). These works, which were mainly developed to address the issue of firms' growth, put forward further empirical evidence on the dual structure the structure of enterprises. The small competitive firms in the fringe can play a diverse role, such as, for instance supplying a surge in demand and a changing variety of oddment for niche markets. The automobile industry has a dual structure both in the vertical integration and in the generation of variety.

Concerning the dualism in variety generation, which is the focus of this book, the oligopolistic core consists in the big three producers plus a few European and Asian importers; as stressed before, they manage to offer a broad range of variety. However, the consumer not yet satisfied by this spectrum of variety can rely on a

[2] Build to order means that a car is assembled in response to a consumer order. It is considered the best way to applied to mass customisation in the automobile industry (Agrawal et al. 2001).

peculiar niche market. The aftermarket for specialty components offers products to customise a vehicle. According to the SEMA[3] definition, this niche market consists *in a unique industry dedicated to the enhanced performance, appearance and handling of all types of passenger cars, light trucks and recreational vehicles* (SEMA 2002).

The tendency to customise cars is as old as the history of the Model T. As explained above, the 20-horsepower engine of the model T was used also for agriculture. Companies like L.A. Tractor Company and the Handy Hank entered the market of specialty components to convert the Model T into a tractor. However this niche market had a short life: first, cars have lost very soon the simplicity in design of the early days. If in those early days was even possible to convert the car into a snow machine, thereafter the complexity of the vehicles increased together with the improvements in the performance. The technical change that made possible the refinements of the product (like the incorporation of electrical wiring or hydraulic power steering, many aerodynamics considerations, and the use of new materials) raised also the difficulty of substantially modifying a vehicle. Secondly, tractors require different characteristics than a vehicle like for instance more horsepower, a different cooling system and more robustness since they have to work continuously at full load. When specialise companies like Ford itself and the Massey Ferguson Farm Tractors started mass producing small tractors soon after the World War I, these make-shift tractors were not competitive enough to survive (Wik 1972, p. 33; Leffingwell 1991).

However, the modern specialty component niche market has a different origin and it is rooted in the 1950s when two groups of consumers started to modify their vehicles. The first were the "customisers" that were reworking the bodylines mainly for aesthetic or artistic reasons. The second were the "hot-rodders", who were more interested in enhancing performance by changing fuel-injection system or adding extra carburettors (Rae 1965, p. 316). Now, however, the specialty equipment market is no longer a niche market for few "hot-rodders", but it rather covers a large market of consumers that want a car more tailored to their needs.

This market, although large, is highly fragmented with 1,524 small or medium size companies. These firms serve a niche market steadily growing in the last 10 years from total sales of 2.35 billion dollars in 1985 to 9.02 billion dollars in 2001 (Table 1.1). By way of comparison, Table 1.2 shows the total sales of other sectors.

Market fragmentation is high also in term of products. There are about 500 basic products available in several variants. In addition, products vary distinctively from producer to producer (SEMA 2002). They can be grouped in three macro areas: accessories and appearance, racing and performance, wheels, tires and suspension (Table 1.3). In these macro-categories the variety of product is very broad: for instance the accessories and appearance categories groups product that

[3] SEMA (Specialty Equipment Market Association) is the trade of association of companies producing specialty equipment.

**Table 1.1** Growth of the specialty components niche market

| Years | Total sales | Change (%) |
|---|---|---|
| 1985 | 2.35 | – |
| 1986 | 2.63 | 1.91 |
| 1987 | 3.07 | 6.73 |
| 1988 | 3.51 | 4.33 |
| 1989 | 3.97 | 3.11 |
| 1990 | 4.35 | 9.57 |
| 1991 | 4.19 | 3.68 |
| 1992 | 4.58 | 9.31 |
| 1993 | 5.14 | 2.23 |
| 1994 | 5.47 | 6.42 |
| 1995 | 5.96 | 8.96 |
| 1996 | 6.32 | 6.04 |
| 1997 | 6.85 | 8.39 |
| 1998 | 7.47 | 9.05 |
| 1999 | 8.17 | 9.40 |
| 2000 | 8.69 | 6.29 |
| 2001 | 9.02 | 3.84 |

*Source* SEMA (2002)

**Table 1.2** Comparison with other sector (sales in billion $)

| Sector | Total sale |
|---|---|
| Athletic footwear | 9.1 |
| Pleasure boats and motors | 8 |
| Exercise equipment | 3.6 |
| Golf (balls and clubs) | 2.5 |
| Bicycles and accessories | 2.5 |
| Camping equipment | 1.6 |
| Baseball/softball | 0.4 |
| Basketball | 0.3 |

*Source* SEMA (2002)

**Table 1.3** Manufacturer sales by segment (sales in billion $)

| Segment | Sales | Share (%) |
|---|---|---|
| Accessories and appearance | 5.12 | 56.76 |
| Racing and performance | 1.71 | 18.96 |
| Wheels, tires, and suspension | 2.19 | 24.28 |
| Total | 9.02 | 100 |

*Source* SEMA (2002)

range from sunroofs to ground effects, hood deflectors, tail light lens covers, intermittent wiper controls, and custom sound systems.

In sum, many historical factors contributed to shape this dual structure and the evolution of variety generation in this industry: the development of new technologies like the internal combustion engine, the choices made by few single individuals such as Henry Ford and Taiiche Ohno, the institutional setting as the case of the Japanese labour legislation, and, finally, the evolving consumers preferences, like the presence of "hot rodders" and "customisers". Since history matters, research about variety in this sector cannot be carried out in vacuum and, therefore, this chapter provided the knowledge required to interpret the empirical results in relation with the historical events. Moreover, this chapter pinpointed the duality in the automobile industry and this property of the structure will be the basis for the methodology used to investigate the data. For an interpretation of empirical results, however, knowledge on the context is not sufficient but a theoretically basis is necessary for a guidance through the data. The next chapter addresses the theoretical basis of an analysis of variety generation.

# Chapter 2
# The Economics of Variety

In this chapter, I review the theoretical basis for the economic analysis of variety. The goal of this chapter is twofold: I want to position critically my work and developed few useful concepts. In particular, I will describe two approaches to the issue of variety and point out their different conclusions on the nature of variety generation and its limits.

## 2.1 Variety Generation in the Product Life Cycle

A first approach stems from Abernathy and Utterback (1975, 1978) who tried to explain some stylised facts about the industrial and technological dynamics in the automobile sector.

In the early days of an industry there is a wave of new entrants whose aim is to take advantages of opportunities to sell various versions of the new artefact: few of them succeed and grow rapidly and many die and are replaced by new entrants. In this first phase, opportunities and turbulence are high, concentration low. Thereafter, the market stabilizes, entry slows, the number of firms reaches its peak and a slow and continuous shake out occurs reducing drastically over time the number of extant firms. Together with this process there is a shift from product innovation to process innovation (Klepper 1996).

Several theoretical reasons have been given for this story. Abernathy and Utterback (1978), provided the first and the most influential explanation. They suggested the idea of a dominant design linked with a technology life cycle underlying the development of the industry.

When a new technology emerges, the environment is characterised by high uncertainty about how the technology can satisfy users' requirements and users themselves are not well aware of their needs. In this phase, there is lot of experimentation and learning, and product innovations are frequent (Burns and Stalker 1966). Moreover, uncertainty about the future development of the industry restrains firms from investing in process technology, which is not easy to redeploy. In this fluid phase in the automobile industry, there were attempts to produce



M. Guerzoni, *Product Variety in Automotive Industry*,
SpringerBriefs in Business, DOI: 10.1007/978-3-319-01907-9_2,

petrol, electric, and steam engines. It was uncertain whether the internal combustion engine would become the standard in the end (Abernathy 1978).

The interplay between depletion of technological opportunities on the firms' side and determination of preferences on the demand side serve to select a standard version of the product, the dominant design in the jargon of Abernathy and Utterback. Dominant design is *a specific path (…), which establishes dominance among competing design paths* (Suarez and Utterback 1995, p. 416).

As described in the model developed by Abernathy and Utterback, when a radical innovation emerges, before the industry finds a satisfactory compromise between form and context, there is a period where different designs are competing (Abernathy and Utterback 1975). Competing on design means to find product attributes that best fit with demand requirements. The Model T Ford did not emerge as dominant design because it was cheaper and more robust than the other cars *ceteris paribus*, but because Henry Ford decided that the product design had to focus on cheapness and robustness, rather than, for instance, aesthetics, comfort, and originality.[1]

Once a dominant design has emerged, only the firms able to produce it efficiently will survive, there are fewer entrants because entry barriers became higher and, therefore a shake out occurs. From the technology side, there is progressive shift from radical to incremental innovation and from product to process innovations. The last phase of the industry is the maturity: only few marginal changes are made on the product, competition is based on price and, therefore, on cost-cutting technologies improvement. The few firms left are those big companies with a large market that allows them to exploit economies of scale. It has been later acknowledged that the distinction between radical and incremental innovation was incomplete in the mature phase of a cycle and the dichotomy between architectural and component innovation was introduced (Henderson Clark 1990). A component innovation involves modification of a part of a sub system, that leaves the existing structure unchanged. On the other hand, *the essence of an architectural innovation is the reconfiguration of an established system to link together existing components in a new way* (Henderson and Clark 1990, p. 11).

The success of the Product Life Cycle (PLC) theory is due to its powerful simplicity in accounting for many stylised facts of industrial dynamics. It can be used by consultants to assess the strategic position of firms[2] and by evolutionary economists who coupled the idea of dominant design with that of technological trajectory (Dosi 1982). Dosi talks about *reciprocal consistency* between Abernathy and Utterback and his own thesis as common.

---

[1] Consumers perceived the car as an strange horseless carriage for rich people. Therefore, competition was based upon comfort as in the most luxuries market segment of carriage industry. On the contrary, Ford offered a new concept of the product (Clark 1985).

[2] For instance the PLC is the underlying concept of the Boston Consulting Group (BCG) Matrix. Stern and Stalk (1998) contains the original contribution by Bruce Henderson on the BCG matrix.

> painstaking attempt to construct a non neoclassical theory of technical change capable of giving a satisfactory account of (1) the relationship between economic forces and the relatively autonomous momentum that technical progress appears to maintain, (2) the role of supply side factor… (Dosi 1982, p. 159).

The conceptual proximity of technological trajectory and dominant design is strengthened because PLC theory is intrinsically a theory of lock-in. Clark explained this characteristic by recalling two properties of design. First, the concept of design, as introduced by Simon (1962) and Alexander (1964) can be defined as solution to a given problem. Alexander calls "form" the solution and "context" the problem. Then, he explained,

> design is not only a matter of form that should be adequate to the user, but every design begins with an effort to achieve fitness between two entities: the form and its context. […] In other words, when we speak of design, the real object of the discussion is not the form alone, but the ensemble comprising the form and the context (Alexander 1964 quoted in Clark 1985, p. 236).

The context is defined by consumer needs, whilst the form consists of the attributes embedded in the product. Secondly, Clark explains that the process of searching for a design involves the analysis and the identification of the components of the form and how they may be organised into architecture. These components have different significance in the product and some proceed logically and temporally before others. This process shapes a system that has hierarchical structure where the form of some subcomponents depends upon decisions taken for others at a higher level: major changes cannot occur without alterations to the global architecture. For instance the design of the combustion chamber followed and depended on previous choices about energy transformation, the adoption of internal combustion and the design for power delivery involving the use of pistons (Clark 1985, p. 241).

In the first phase the interplay between form and context is crucial for the process of design, but, after a dominant design has emerged, the form became rigid as consequences of its hierarchical structure. Changes are likely to be incremental within the last level of the hierarchy. Moreover, as time goes by, the hierarchical structure became more branched and incremental change results to be more and more marginal. This is exactly the case of the car that in the beginning it was even possible to convert into a tractor. As put forward by Cebon et al. (2001), this evidence is explained by PLC theory by considering the increasing synergetic specificity both among product's components and between demand and technology. In the case of the car the synergies among components increased and it has been more and more difficult to add parts not designed to fit exactly a car or, even, a specific model. The synergies between demand and technology increased as well: in the early days of the automobile industry, different groups of consumers, like the farmers could benefit from the new product. Thereafter, the design developed specifically to fit the context of car as a passenger vehicle.

Only when architectural innovation occurs, are modifications to the core component of the artefact possible. Always Cebon et al. (2001) suggested that the introduction of modularity could be considered as an architectural innovation that decrease the synergetic specificity among components and reduces the lock-in effect of the dominant design. If, as seen in the previous chapter, different modules can be added on platform, the range of variety increases. A modification of the single module is easier because it does not involve modification of the core components in the platform.

Though powerful, the PLC theory has some weaknesses. In particular the basic version of the PLC underestimates the role of demand. Nelson (1994) and Porter (1990) doubt that a dominant design can emerge in markets whit a heterogeneous demand. Teubal as well as Foray and Gruebler suggest that bifurcation in design trajectories can occur if firms face segmented demand (Teubal 1979; Foray and Gruebler 1990). However the case of automobile has been taken as paradigmatic for PLC theories, which manages to explain the evolution of this industry.

The implications for the generation of variety in PLC theories are straightforward. Variety, in term of different variants of competing products, reaches the peak in the first fluid phase before the dominant design emerges. Thereafter variety is generated by incremental improvements and refinements in the product that follow a relatively technological trajectory. Secondly, these improvements and refinements generate variety only to a certain extent since they involve only components at the lower lever of the design hierarchy. More variety can emerge only if there are architectural innovations that change the structure of the design (Abernathy and Clark 1990). This means that limits in the expansion of variety are endogenous to the product:

Proposition 1: There exist limits in the expansion of variety endogenous to the product.

## 2.2 The Equilibrium Approach to Variety Generation

A theory of generation of variety, *strictu* sensu, in the PLC theory was only an implication of basic considerations about product and technological evolution. Other non-PLC approaches attempt to construct a theory of optimal variety generation. The term product variety is used in this approach *to refer to the number of variants within a specific product group, corresponding broadly to the number of* "brands" *as the term is used in marketing literature or the number of* "models" *in consumer durable market*s (Lancaster 1991, p. 189).

Two bodies of literature are reviewed: models with representative agent and address models. In economics, the theory of variety began as a by-product of the analysis of the deviation from the competitive model (Polo 1993, p. 29). As capitalism evolved, shortcomings in the model of perfect competition became more apparent. A theoretical attack to the Marshallian[3] framework first took place

[3] Marshallian framework refers to *market economies in which every relevant good is traded in a market at publicly know prices and all agents act as price takers* (Mas-Colell et al. 1995, p. 307).

in Cambridge during the 1930s, where Piero Sraffa, dissatisfied with the partial equilibrium analysis, in his article "The law of returns under competitive conditions" (Sraffa 1926), decided to shift its research interests away from perfect competition. Similarly, Edward Chamberlain and Joan Robinson analysed oligopolistic industries under the assumption of firms as price-setting entities (Chamberlain 1933; Robinson 1933). In particular, Chamberlain addresses the issue of product differentiation in oligopolistic industries. In his basic work, he describes a model where preferences are referred to the set of all goods. The classical assumption of quasi-concave utility function attributes to the consumer a taste for variety, in the sense that all goods are purchased by each consumer and her utility is higher, the higher is number of products.[4] Demand is homogenous and can be summarised by a representative agent. From the supply side, each firm is a monopolist producer of a distinct good facing a downward sloping aggregate demand curve. However, it is not monopolistic at industry level; an industry is defined as a group of firms with similar cost structure and whose products have a degree of substitutability. The presence of positive profits in the industry, in the absence of entry barriers, pulls new firms to enter the industry with a variant of the product. This event decreases the residual demands for incumbent firms. When the demand is tangent with the average cost curve, price is equal to marginal cost and marginal revenue is equal to marginal cost: the profit for firms and thus for industry is zero, thus entry stops and there is equilibrium. The outcome of this model is that generation of variety is lowered by reduction in the degree of product substitutability, by increased in the fixed cost of production, and by decrease in the market size. Many criticisms can be made about these approaches, including the inappropriate adoption of the concept of the representative agent (Kirman 1989). However one result is important to retain for the following analysis. The generation of variety requires two necessary conditions that agents show a taste for variety and that firms make profit in producing more variety.

Address models, on the other hand, date back to the seminal paper by Hotelling (1929), "Stability in Competition". The assumptions are rather different from the representative agent model: they assume preferences on the different variants of the good. Consumers can rank them and will buy only the one they perceive as the best. They have heterogeneous preferences and, therefore, rank different varieties of products differently. In order to depict these assumptions, Hotelling describes a model where firms compete on more than one characteristic: for instance, in his paper, they compete on location and price. The model introduced a single dimensional space, a street for instance, where firms can locate to sell their products. The products, although homogenous, differ for the consumer in terms of

---

[4] Quasi-concavity in the utility functions implies that a preference relation on a consumption set is convex. That means that, if the commodity bundle x is indifferent to y, any linear combination of x and y, cannot be worse than either one or the other alone, i.e. a mixture (more variety) is almost preferred. A classic reference is Mas-Colell et al. (1995), Chap. 3. In this body of literature, the most popular contributions are Dixit and Stiglitz (1977), Hart(1985) and Perloff and Salop (1985).

location. Consumers are uniformly distributed along the street and they have to bear transport cost to reach the location of a firm. Under this framework Hotelling state the principle of minimal differentiation, that is that two symmetric firms locate in the centre. It was shown after decades (D'Aspremont et al. 1979) that Hotelling's result was wrong and that the minimal differentiation can be the outcome of a location process only under certain assumptions. However, the importance in Hotelling's contribution was the suggestion of a new way of addressing the problem of product differentiation. Lancaster was the first to develop the idea of Hotelling that the spatial approach can be extended to product differentiation along other dimensions. Lancaster (1966, 1971) developed models where a good can be considered as a "bundle of characteristics" and these characteristics became the unit of the analysis. Consumers express their preference on characteristics, not on the good.

Qualitatively these models have the same outcome as the monopolistic competitions' models: variety depends negatively by substitutability and fixed cost. However, they add an important feature. If in Chamberlain's model a larger market predicts an increase in variety, this outcome is ambiguous in address models. In these models there is a distinction between the width of the market, that is the dispersion of consumers' preferences, and the depth of the market, or the density of consumers at each location (Lancaster 1990, p. 200): therefore a market could be larger but if it is homogenous with consumers concentrated in a certain location, that leads to clustering of firms in this location rather than an increase in variety.

A comparison between these models and PLC models makes little sense because they originate in different economic traditions, aim to answer different questions, and use different methodological tools. However, the reason why I compare them is that they reach a different conclusion on the nature and on the constraints in variety generation.

As seen before, the variety in PLC is mainly considered as improvement in the design and its limits are established by boundaries in the hierarchical structure of the product. The generation of variety is incremental and continuous only to the extent it involves the lower levels of hierarchy and along autonomous technological trajectories. Further improvements can take place only with important and discrete change in the architecture. On the contrary, in the Lancastrian approach, the product does not show any limit in the variety it can achieve. Its form is continuous and flexible; the main street of Hotelling's model is a line infinitely divisible. Only the size and composition of demand, by dictating incentives, and opportunities for development of process technology and thereby production costs, set boundaries to expansion of variety.

Proposition 2: Only the size and composition of demand, by dictating incentives, and opportunities for development of process technology and thereby production costs, set boundaries to expansion of variety.

Since this book is an empirical study of the variety in an industry, it is closer in the epistemological foundation and in the tools it is using to the PLC tradition. However, as explained in the next chapter, both because the methodology roots in the characteristic's approach and because it observes variety at a specific point in time there is also a flavour of Lancastrian models. Nevertheless, since the book aims to increase the understanding on the generation of variety, no assumptions should be made a priori on its nature and on the nature of these limits. On the contrary this dichotomy will drive the empirically analysis and the validation of one of the two theory will be an outcome of the analysis.

# Chapter 3
# Empirical Analysis

## 3.1 Methodology

The aim of this book is to shed some light on the process of variety generation in the US automobile industry. The existence of the peculiar niche market of specialty components for vehicles allows the development a new techniques for the analysis as explained in following paragraphs.

In the literature there are few empirical studies of niche markets. Ecological theory has been addressing the issue of niches for decades and some recent works (Hannan 1997; Sorenson 2000) have tried to extend the result to the dynamics of industrial organization. In particular, the resource partioning theory (Carroll 1985; Carrol and Hannan 1995) seems a promising tool. It tries to explain why mature industries move to the phase of de-maturity (Abernathy et al. 1983) where they experience new entries and the creation of a niche market, defining specialist producers who can develop and provide variety for sophisticated demand not satisfied by the mass producer, the generalist. According to Boone et al. (2002) the explanation for this tendency is due to the process of evolution of consumer's preferences that became more and more sophisticated. In particular, in this stream of literature, the work by Dobrev et al. (2002) "The evolution of organizational niches: US automobile manufacturers, 1885–1981" must be quoted because it addresses the problem of market segmentation and product's differentiation specifically among car producers in the US. That paper recalls the ecological theories that try to unify different assumptions on the relation between niche width and market scale (Dobrev et al. 2001; Dobrev and Carroll, 2000) and, thereafter, performs an historical analysis of the market. However the data at my disposal does not suit their empirical method: they used a long time series that considers only one product characteristic; in this book, data show many characteristics at a single point in time. Moreover in their case the niche market is producing variants of the product. Here the niche market provides the consumer with the possibility of changing the product herself. Other works, with an evolutionary flavour, approached the study of bifurcation, niche creation and the role of niche markets in the evolution of technologies (Frenken et al. 1996). These works address the evolution

M. Guerzoni, *Product Variety in Automotive Industry*,
SpringerBriefs in Business, DOI: 10.1007/978-3-319-01907-9_3,

of industry by measuring the distance between products in a characteristics space *à la* Lancaster. Again and for the same reasons cited above, the peculiar characteristics of the niche market in this analysis are not tailored for this empirical method either.

However, the originality of the case analysed in the book should not be perceived as a weakness. On the contrary, it allows the development of a new technique to improve the understanding of variety's generation: the presence of this peculiar niche market allows a comparison between the variety provided in the market with the possibility of its modification through the niche market. This methodology can be defined as "niche market analysis". Its originality does not lie in the analysis of the niche market *per se*, but because niche market is used as tool for examining the boundary of variety generation of the primary producer. This methodology relies on the assumption that this niche market can be taken as a proxy for the unfulfilled demand for variety of the consumers both in term of size and composition.

Other assumptions could be made. For instance, it could be argued that consumers show an absolute preference for niches, that is, that they buy in a niche market and customised their vehicle rather than buying it from large companies, independently form the degree of customisation those offer. Two explanations for this behaviour probably coexist in each "customisers". A consumer of the niche market might want to customise the car either because of a demand for distinction,[1] or just as a hobby in the free time and not because of a unfulfilled demand for variety. However, if it does make any sense looking for rationality in human vanity, it is reasonable to assume that, if everyone could easily access to a perfectly customise car, consumers will try to satisfy their demand for distinction in a different way. Given, thereafter, that the car is far to be a perfectly customisable good, it is reasonable to assume as well that consumers will try to customise their car by modifying those characteristics where the supply of variety by large companies is most limited. Therefore, it is possible to address the issue of the limits in variety's generation by highlighting the variety that large companies are not able to generate, but that is required because consumers look for it in the niche market. The niche market analysis is a three steps process.

*Step 1. Variety generation among car producers*. The first step is to analyse the variety generated by the large car producers. Ford is taken as a proxy for the product differentiation strategy of all the large companies in the United States. This assumption is reasonable since Ford is selling cars in all of the market segments defined by the Nada (National Auto Dealer Association). Moreover Ford is paradigmatic for the evolution of the variety strategy: from the only black Model T in 1908 to the *build-to-order* strategy in 2003. In order to analyse Ford's product differentiation I follow the characteristic approach *à la* Lancaster. The empirical

1 "demand for distinction" means that consumer seek a unique product to emphasise the distinction from other people. See Swann (2001) an empirical case and Shy (2001) for a theoretical approach.

use of the Lancastrian idea of a product as a bundle of characteristic has a long tradition (Saviotti and Metcalfe 1984; Saviotti 1996; Frenken and Leydesdorff 2000). Saviotti and other scholars analysed the evolution of variety by calculating diversity's indexes like the entropy index or the Weitzman's measure[2] for several product's characteristics over time. Although the structure of the data, with mainly dummy variables, does not allow the implementation of this kind of indexes and techniques, the characteristics approach can be still used.

Moreover Saviotti's approach goes further and considers two types of characteristics: technical characteristics and service characteristics. Technical characteristics are defined as variables that can directly be manipulated by producers (e.g., engine power). Variables that users take into account in their purchasing decisions (e.g., speed) have been considered as service characteristics. (Frenken and Leydesdorff 2000, p.3). Saviotti and his scholars operated this division because they need to model the supply side as a set of technical characteristics in order to depict change in design as a change in the interfaces between demand and supply (Frenken Leydesdorff 2000 and Simon 1962 quoted in Frenken Leydesdorff 2000). Since the specialty components niche market provides the opportunity to modify not only service characteristics (speed, colours, etc.) but also technical ones (internal characteristic of engine or gear box, wheels), it is not useful to operate the same distinction. A pure Lancastrian approach has therefore been chosen.

However, I postulate a different dichotomy, which is not assumed before the analysis, but it will be one of the outcomes of the analysis. I try to distinguish between characteristics whose variety is set ex ante by the firm (variety as differentiation) and those that can be chosen by the client at the moment of purchasing a car (variety as customization). Since, as explained, the niche market provides the opportunity to customise the product, this distinction is crucial in order to compare the two markets.

*Step 2. Variety generation in the niche market*. The second step of a niche market analysis is to understand the type of variety offered by the niche market. The available data provide, for each firm in the niche market, information on the type of product it is selling and assign each product to a product group. A correspondence needs to be developed between product's groups and the characteristics of the vehicle they can modify. In this way, the analysis becomes commensurable with one conducted on variety in Ford. In this second step, it is also useful to understand how the division of labour of variety's generation is organised within the niche market. That will help explain why the large producers do not generate this variety.

*Step 3. Comparison*. The last step consists in the comparison of the variety generated by Ford and by the group of niche producers. What it is required is not to

---

2 Entropy index is a concentration index that derives from information theory (Theil 1972). Weitzmann's measure (Weitzmann 1992) is a maximum likelihood grouping techniques that measure the degree of dissimilarity between two elements according to a measure of distance of their characteristics.

compare the degree of variety in the industry as represented by Ford with the degree in the niche market for each characteristic. The aim of the comparison is rather to understand where in the space of characteristics that car producers are offering the largest variety (e.g. number of colours) and where the niche producers offer the highest possibility for modification (e.g. type of wheel). Then the process of analysis should be integrated with the knowledge of history of the sector to put forward a convincing explanation of the empirical results.

## 3.2 Variety Generation at Ford

### *3.2.1 Data*

The first dataset provides information on all the models sold by Ford in 2003. By browsing NADA website for each trim of the 19 Models is possible to access product's characteristics. In total there are 344 trims. Trims are versions of the model that differ by some key characteristic such as type of engine, horsepower and availability of optional items. The models are grouped in five categories according to the market segment that they address (passenger, luxuries-passenger, mini-vans and vans, sport utility vehicles, pick-up trucks.). Table A1 in the appendix resumes the cases in the dataset.

For each trim NADA website provides 79 characteristics that define the vehicle. In every case, NADA organised the characteristic in nine groups (price, specification, safety, comfort, entertainment, performance, interior, exterior, warranty). All of the variables are in the data set, but, in the analysis, some of them were dropped for different reasons: some are tautological repetition of other variables like, for instance, the base MSRP,[3] price and the base invoice price. Some others are not interesting for the analysis either because they are optional items not available in any trim (e.g. dual climate control system) or they are standard for all of them (e.g. power windows). Others have been dropped because they are not characteristics of the car but services provided with it (like the warranties) and, finally, some others had too many missing values. At the end, the analysis is going to be restricted to 44 characteristics. All of the characteristics from the group warranty were dropped. The following tables report results also for price, but it is not part of the analysis, because it is not a characteristic of the artefact. At the end, therefore the groups in the analysis are seven. Table A3 in the appendix summarises the dataset. It specifies the variables in the analysis together with the categories to which they are assigned, whether they are used in the analysis and, if not, the reason why are not used.

---

[3] Minimum Recommended Selling price.

### 3.2.2 Data Analysis

To have a first grasp of the product differentiation, I calculated the coefficient of variation (CV) for each characteristic. Given a distribution, the coefficient of variation is its standard deviation divided by the absolute value of its mean:

$$CV_j = \frac{\sigma_j}{|\mu_j|} j = 1, \ldots 44 \tag{3.1}$$

It is an index of dispersion; the higher it is, the higher is the variety available for the single characteristic. There was not any particular reason to correct the standard deviation and, therefore, I used the population standard deviation. A property of the coefficient of variation is that it is a pure number and it makes possible a comparison among variables measured with different units, as is the case of this dataset. For those characteristics measured with a cardinal unit the coefficient of variation has been directly calculated using the value of the variable.

For binary and multinomial variables a slightly different method has been followed. Binary variables explain whether a characteristic can be added or not to the trim (for instance a consumer can decide to add a CD changer or not), multinomial variables specifies between a set of variants which one is applied (type of engine). For binary variables, the coefficient of variation was calculated on the probability that a characteristic can be added to the trim and, for multinomial ones, on the probability that a given variant of the characteristic is used. The probability here is observed as relative frequency. Table 3.1 ordered the characteristics according to the coefficient of variation. The group each characteristic belongs to is also specified in Table 3.1.

It is possible to notice that:

1. There is an uneven degree of variety;
2. there is a strong correlation between group of characteristic involved and degree of variety.

In particular, characteristics defining exterior, performance, interior and specification features have a low coefficient of variation while entertainment, comfort, and safety show a higher degree of variety (Table 3.3 and Graph 3.1(Fig 3.1)). The exterior and interior features define the physical dimensions of the vehicle. The predominant contemporary production process still involved pounding roles of steel sheets against dies. A die is a piece of metal with the shape that the steel will assume under pounding. Therefore variation in this part implies investments in heavy and expensive process technology. Variety in safety, comfort and entertainment depends on the range of optional items that can be added. Therefore it is not surprising that these groups show the highest degree of variance.

**Table 3.1** Coefficients of variation per characteristic

| Group of characteristic | Characteristic | CV |
|---|---|---|
| Interior | Front head room | 0.03 |
| Interior | Front leg room | 0.03 |
| Interior | Second row head room | 0.04 |
| Exterior | Length | 0.07 |
| Exterior | Width, max w/o Mirrors | 0.09 |
| Interior | Front shoulder room | 0.09 |
| Interior | Second row leg room | 0.10 |
| Interior | Front Hip Room | 0.11 |
| Exterior | Height–overall | 0.12 |
| Exterior | Wheelbase | 0.16 |
| Specification | Transmission type | 0.16 |
| Interior | Second row shoulder room | 0.21 |
| Interior | Second row hip room | 0.21 |
| Performance | Power | 0.22 |
| Comfort | Cruise control | 0.24 |
| Price | Price | 0.27 |
| Performance | Distribution hitch | 0.28 |
| Interior | Passenger capacity | 0.33 |
| Performance | Torque | 0.33 |
| Specification | Fuel Tank capacity gals | 0.34 |
| Performance | Deadweight hitch max tongue | 0.41 |
| Comfort | Remote keyless entry | 0.43 |
| Performance | Dead weight hitch-max trailer | 0.45 |
| Entertainment | Cassette player | 0.52 |
| Entertainment | CD changer | 0.65 |
| Specification | Drive train | 0.72 |
| Interior | Cargo volume | 0.72 |
| Comfort | Power Seats–driver | 0.75 |
| Exterior | Colour combination | 0.90 |
| Comfort | Leather seats | 0.96 |
| Safety | Headlights–automatically on | 1.01 |
| Specification | Engine type | 1.05 |
| Interior | Rear window defroster | 1.12 |
| Interior | Rear tires | 1.21 |
| Safety | Child safety locks | 1.25 |
| Interior | Sunroof/moon roof | 1.38 |
| Safety | Air bag–passenger switch | 1.42 |
| Comfort | Vehicle anti-theft protection | 2.02 |
| Safety | Traction control | 2.18 |
| Comfort | Automatic climate control | 2.37 |
| Entertainment | CD player | 2.40 |
| Safety | Air bag–side–body–front | 2.49 |
| Comfort | Universal garage door opener | 2.58 |
| Safety | Air bag-side–head only–front | 2.69 |

*Data Source* Nada Website. My elaboration

**Table 3.2** Average coefficients of variation per group of characteristics

| Group of characteristic | Average CV |
|---|---|
| Exterior | 0.27 |
| Performance | 0.34 |
| Interior | 0.43 |
| Specification | 0.57 |
| Entertainment | 1.19 |
| Comfort | 1.34 |
| Safety | 1.84 |

*Source* Nada. My Elaboration

### *3.2.3 Market Segmentation and Model Customisation*

In order to refine the analysis, I have to make same working hypotheses on the way the product's variety is organised in the market.

The working hypothesis used here it is that a customers wants to buy a car either for a specific purpose or for idiosyncratic preferences for a model or category. This is a reasonable hypothesis because the car is a mature product and both consumers and producers have developed during decades their consumption and production capabilities (von Tunzelman et al. 2003). The firm is aware of the market segmentation and, therefore, produces different model's designs to fit best each segment.

Once a model is chosen, the producer offers the consumer the possibility to further customise the car in two ways: within model, she can choose different trims and, thereafter she can customise the vehicles as much as possible to meet her needs by adding special features. There exist, therefore, two different types of product's variety: the first one is set ex ante by the producer (variety as segmentation), the second is more flexible and allows the consumer to make choices[4] (variety as customization).

I try now to distinguish among the characteristics whose variety has been set *ex ante* by the firm and those that can be customised by the consumer according to her tastes.

This is an important step in the empirical analysis since, as expressed in the methodological discussion, I want to compare the variety in the industry with the possibility of "do it yourself" offered by the niche market, which impacts on the variety as customization. It is therefore crucial, to highlight the real

[4] It could be argued that a consumer does not choose a model but a category. Consumers tend to choose a very narrow market segment. Therefore, they compare different brands in the same small segment rather than different models from the same brands (for an introduction to consumer choice see Armstrong and Kotler 2003). Data are here available only for one brand. Therefore, the consumer category collapses to one model.

**Table 3.3** M index for characteristics

| Type of Characteristic | Characteristic | M |
|---|---|---|
| Exterior | Width, max w/o mirrors | 0 |
| Exterior | Length | 0 |
| Interior | Front shoulder room | 0 |
| Interior | Front head room | 0 |
| Interior | Front hip room | 0 |
| Interior | Front leg room | 0 |
| Interior | Second row shoulder room | 0.01 |
| Interior | Second row head room | 0.01 |
| Interior | Second row hip room | 0.01 |
| Exterior | Height–overall | 0.02 |
| Exterior | Wheelbase | 0.03 |
| Interior | Second row leg room | 0.03 |
| Performance | Distribution hitch | 0.04 |
| Performance | Dead weight hitch-max tongue | 0.05 |
| Specification | Fuel tank capacity gals. | 0.06 |
| Performance | Dead weight hitch-max trailer | 0.07 |
| Interior | Cargo volume | 0.08 |
| Performance | Torque | 0.08 |
| Specification | Drive train | 0.08 |
| Performance | Power | 0.09 |
| Interior | Passenger capacity | 0.11 |
| Price | Price | 0.12 |
| Comfort | Cruise control | 0.17 |
| Specification | Transmission type | 0.19 |
| Exterior | Colour combination | 0.23 |
| Specification | Engine type | 0.23 |
| Exterior | Rear tires | 0.29 |
| Comfort | Remote keyless entry | 0.35 |
| Entertainment | CD changer | 0.44 |
| Entertainment | Cassette player | 0.5 |
| Comfort | Power seats–driver | 0.55 |
| Comfort | Leather seats | 0.56 |
| Comfort | Rear window defroster | 0.79 |
| Comfort | Headlights–automatically on | 0.91 |
| Comfort | Sunroof/moon roof | 0.97 |
| Safety | Child safety locks | 0.97 |
| Safety | Traction control | 1.29 |
| Comfort | Automatic climate control | 1.44 |
| Entertainment | CD player | 1.49 |
| Safety | Air bag–side–body–front | 1.56 |
| Comfort | Vehicle anti-theft protection | 1.62 |
| Safety | Air bag–passenger switch (on/off) | 1.64 |
| Comfort | Universal garage door opener | 1.81 |
| Safety | Air bag–side–head only–front | 3.18 |

*Source* NADA website. My elaboration

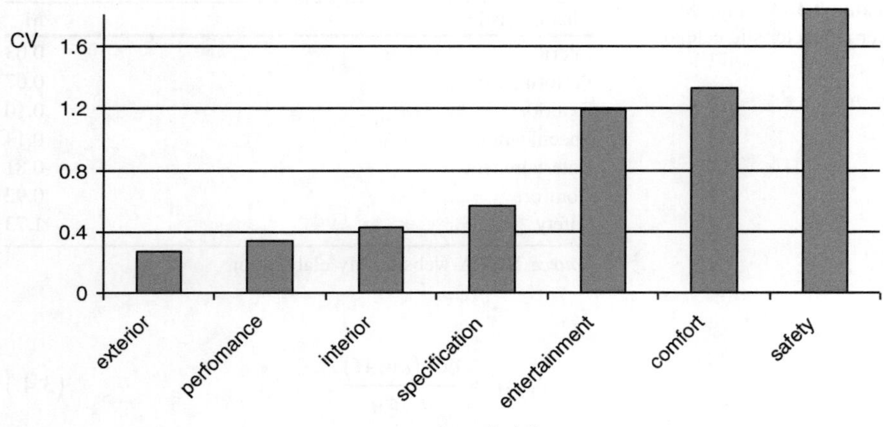


**Graph 3.1** Coefficient of variation per group of characteristics. *Source* Nada. My Elaboration (Table 3.2)

customisation implemented by the consumer by adding optional items, choosing colours, etc. and to distinguish it from the market segmentation.

In order to pursue this task, I first calculated for each characteristic *j* the coefficient of variation within each model *i*, (*CVij*), and I considered its average across model $M(CV)_{ij}$ as proxy of the degree of freedom in customising a vehicle once a model has been chosen:

$$CV_{ij} = \frac{\sigma_{ij}}{|\mu_{ij}|} \quad \text{where } i = 1,\ldots,19 \text{ and } j = 1,\ldots,44 \tag{3.2.}$$

$$M(CV)_j = \frac{\sum_{i=1}^{19} CV_{ij}}{19} \tag{3.3.}$$

The following tables depict the value of the index:

If after having chosen a model, the customer faces lot of variety, i.e. a high average coefficient of variation per model, there is something similar to a process of customisation for that characteristic.

There is again a sharp distinction between characteristics that can be customised and those which cannot (Table 3.4). For instance the characteristics define the physical properties of the car do not show little variety within the trim. It is not surprising that for each characteristic the coefficient of variation calculated across all the trims is higher than the average of the within-model coefficients. Trims are version of the model and thereafter they show common characteristic and this lowers the variability. On the contrary, it is worth noting, that the correlation between the two coefficients of variation, $\rho$, is extremely high (0.91).

**Table 3.4** Average M index per characteristic category

| Characteristic | M |
|---|---|
| Interior | 0.03 |
| Performance | 0.07 |
| Exterior | 0.10 |
| Specification | 0.14 |
| Entertainment | 0.81 |
| Comfort | 0.92 |
| Safety | 1.73 |

*Source* NADA website. My elaboration

$$\rho_{cv,M} = \frac{\text{cov}(cv, M)}{\sigma_{cv}\sigma_M} \tag{3.4.}$$

Cov is the covariance between the arrays of the two indexes and $\sigma$ their standard deviation. This has an important meaning and a consequence. It implies that, for each model Ford tends to put on disposal all the variety it can generate in each characteristics. Ford does not leave some optional features or additional modules for a particular model and some others for a different one. If a characteristic can be customised, this possibility tends to be available in all of the models.

However this high correlation proves that some characteristics show a natural tendency towards variety because they are more flexible than others both at level of the sample and within the model. Therefore this coefficient of variation does not help to distinguish among characteristics differentiated by the firm and customisable ones. It is not possible to distinguish whether M has a high value, for exterior if you like, because you can customised those characteristics after having chosen the model or, rather because those characteristic are really flexible both in term of market variety as segmentation and variety as customisation. For this reason M is not very effective in distinguishing among characteristics set ex ante by the firm and customisable ones.

Therefore, I have developed an index (CUST) to distinguish between segmentation and customisation. It is the ratio between the two coefficients of variation taken into account before; it is thus a pure number as well and could be used to compare different types of variables. The numerator is the average of the 19 coefficients of variation calculated within each model. It describes the variety, on average, of the specific characteristics once a model has been chosen. The denominator is the coefficient of variation of the distribution of the characteristics across all trims. The denominator tells, in absolute terms, to what extent a characteristic shows variety as explored in the previous paragraph. Given this absolute degree of variety, the index explores whether a characteristic's variety is grouped in the trim or is evenly distributed.

$$CUST_j = \frac{M(CV)_j}{CV_j} \quad \text{where}\, j = 1, \ldots, 44 \tag{3.5.}$$

This index shows interesting properties. It ranges from zero to positive infinite. It tends to zero when there is a poor variety within the model compared with the variety across the trims. It means that the variety of the characteristic is set ex ante by the firm more than being a possibility of customisation. On the contrary, the index increases, when trims within a model differ more than the models do. This implies that a characteristic is relatively highly customisable once a model has been chosen. *A caveat* should be made. This index underestimates customisation for characteristics with high degree of variety and overestimates it for rigid characteristics. As the following analysis shows, this does not result to be a problem but on the contrary it will confirm the findings a *fortiori*. Table 3.5 shows the variable ranked by the index. A dichotomy in the variables clearly emerges.

A group of variables comprising those defining interior and exterior characteristics has a small value: the highest value is 0.33 for passenger capacity, the average value for interior characteristic is 0.13, and that for exterior is 0.15 (Table 3.6).

On the other hand, characteristics defining comfort, safety, and entertainment show higher values. The smallest value is 0.58 for leather seats, the average value for comfort is 0.73, the one for entertainment is 0.76, and the one for safety is 0.87. Moreover, this group of variables showed also a high coefficient of variation. For this reason, as explained before, the results are true a fortiori because the index should underestimate their degree of customisation.

Both the description of the historical evolution of variety and the theoretical part considered modularity a key factor in the process of customization. The final analysis carried out is the investigation of the role of modularity. I considered modular those characteristics measured by a binary variable. Their binary nature captures the fact that this characteristic are modules that can be added or not to the vehicle. As Table 3.7 show modular characteristics, have a much higher value of the index CUST and confirm the hypothesis put forward in the theoretical part.

In sum, there is a clear dichotomy in the characteristics. On the on side some of them show a high coefficient of variation, they are highly customisable and have a binary nature because they are modules. The others are physical characteristics of the car, and are hardly modifiable. Not only should these parts be modified before the production, but even in the case of *just-in-time* or *build-to-order* car producers can not supply great variety because of the high amounts of fixed investments required for the press. The variation in these characteristics is not a matter of customisation; it is decided ex ante by the firm through the process of market segmentation.

The analysis elucidates the way the company organizes product variety: some product's characteristics define a platform and, thereafter, many modular features can be added on it. Because the platform is not flexible, firms, should set the differentiation, as part of the market segmentation decision. Other characteristics, typically the modularised ones, show a higher degree of customisation. These characteristics concern comfort, entertainment and safety.

**Table 3.5** CUST index

| Type of variable | Type of characteristic | Characteristic | CUST |
|---|---|---|---|
| Cardinal | Interior | Front shoulder room | 0.02 |
| Cardinal | Interior | Front hip room | 0.02 |
| Cardinal | Exterior | Width, max w/o mirrors | 0.03 |
| Cardinal | Interior | Second row shoulder room | 0.03 |
| Cardinal | Exterior | Length in. | 0.04 |
| Cardinal | Interior | Second row hip room | 0.06 |
| Cardinal | Specification | Drive train | 0.06 |
| Cardinal | Interior | Front head room | 0.09 |
| Cardinal | Performance | Dead weight hitch | 0.11 |
| Cardinal | Interior | Cargo volume | 0.12 |
| Cardinal | Exterior | Height−overall in | 0.13 |
| Cardinal | Performance | Dead weight hitch max trailer | 0.15 |
| Cardinal | Interior | Front leg room | 0.16 |
| Cardinal | Performance | Distribution hitch max trailer | 0.16 |
| Cardinal | Specification | Fuel tank capacity | 0.17 |
| Cardinal | Exterior | Wheelbase | 0.17 |
| Cardinal | Interior | Second row head room | 0.18 |
| Multinomial | Specification | Engine type | 0.21 |
| Multinomial | Exterior | Rear tires | 0.24 |
| Cardinal | Performance | Torque | 0.25 |
| Cardinal | Interior | Second row leg room | 0.26 |
| Cardinal | Exterior | Colour combination | 0.26 |
| Cardinal | Interior | Passenger capacity | 0.33 |
| Cardinal | Performance | Horsepower | 0.42 |
| Cardinal | Price | Price | 0.45 |
| Binary | Comfort | Leather seats | 0.58 |
| Binary | Safety | Traction control | 0.59 |
| Binary | Comfort | Automatic climate control | 0.61 |
| Binary | Entertainment | CD player | 0.62 |
| Binary | Safety | Air bag−side−body−front | 0.63 |
| Binary | Entertainment | CD changer | 0.68 |
| Binary | Comfort | Universal garage door opener | 0.70 |
| Binary | Comfort | Sunroof/moon roof | 0.70 |
| Binary | Comfort | Rear window defroster | 0.70 |
| Binary | Comfort | Cruise control | 0.70 |
| Binary | Comfort | Power seats−driver | 0.74 |
| Binary | Safety | Child safety locks | 0.78 |
| Binary | Comfort | Vehicle anti-theft protection | 0.80 |
| Binary | Comfort | Remote keyless entry | 0.81 |
| Binary | Comfort | Headlights−automatically on | 0.90 |
| Binary | Entertainment | Cassette player | 0.97 |
| Binary | Safety | Air bag−passenger switch (On/Off) | 1.15 |
| Multinomial | Specification | Transmission type | 1.16 |
| Binary | Safety | Air bag−side−head only−front | 1.18 |

*Source* NADA website. My elaboration

**Table 3.6** Average CUST index per type of characteristic

| Type of characteristics | Index average |
|---|---|
| Interior | 0.13 |
| Exterior | 0.15 |
| Performance | 0.22 |
| Specification | 0.40 |
| Price | 0.45 |
| Comfort | 0.73 |
| Entertainment | 0.76 |
| Safety | 0.87 |

Source: NADA website. My elaboration

**Table 3.7** Average CUST for type of variable

| Type of variable | CUST |
|---|---|
| Binary | 0.77 |
| Multinomial | 0.54 |
| Cardinal | 0.16 |

*Source* NADA website. My elaboration

## 3.3 Variety Generation in the Niche Market

### *3.3.1 Data*

The second data set has been built with information available on the SEMA (Specialty Equipment Market Association) websites updated to December 2003. SEMA is the trade association of American firms operating in the automotive equipment industry. From its website, it is possible to browse among products and download information on the firms selling the products. This process has been done for all the 412 products, building a database with 1,513 firms. It is now possible to say which firms are selling which products. Products are grouped by SEMA in 20 homogenous categories.

In order to make this data set commensurable with the former and perform a comparison, a correspondence between the products sold in the niche market and the characteristics that they modify in the car has been created. The 20 products' groups are classified in the same eight categories that have been used to define the characteristics of cars in the analysis of variety's generation at Ford. This was possible because the products sold in the niche market are very specific and it is easy to understand their use.

This correspondence was created mainly following the market segmentation of SEMA but in one case it was not straightforward. This was the case of products for engine. The classification was too broad and for some products, it was not clear whether they should have been grouped in the category "performance" or "specification". The borderline between performance and specification is very

**Table 3.8** Number of products per category

| Category | Number of products |
|---|---|
| Accessories for exterior | 90 |
| Chassis, suspension and brake product | 50 |
| Accessories for interior | 30 |
| Drive train product | 28 |
| Mobile electronic | 25 |
| Intake and fuel system product | 24 |
| Product for engine (internal) | 21 |
| Safety equipment | 21 |
| Temperature control product | 16 |
| Engine product (external) | 15 |
| Accessories for tires | 15 |
| Electrical product for engine | 13 |
| Electrical product: lightening | 18 |
| Tire | 12 |
| Accessories for wheels | 10 |
| Wheel | 9 |
| Exhaust system product | 9 |
| Engine (entire or blocks) | 3 |
| Electrical product: security | 2 |
| Component vehicles | 1 |
| Total | 412 |
| Average | 21 |
| Standard deviation | 19.36 |
| Coefficient of Variation | 0.92 |

*Source* SEMA. My elaboration

narrow. Performance and specification obviously are highly correlated since performance depends on the type of engine and other physical characteristics of the vehicle. As a rule, I grouped products for engine in the "specification" category where they had a deep impact in the internal mechanism of the car (for instance new piston). Products were grouped under performance when the enhancement in performance involved a more superficial modification. Following the exposed rationale, products either for modifying or changing internal parts of the engine (like piston, cylinder heads, and crankshafts) were considered products that changed the specifications. On the contrary, products that modify some external part (like motor mounts) were considered as products enhancing performance.[5]

At the end, the 20 groups of products are organised in the eight categories of characteristics specification. No product falls in the category price and warranty, but also in the previous analysis these two categories have been dropped. The categories comfort and entertainment are considered together because the

[5] Technical information derived both from Advance Auto Parts' Website which provide a list of possible modification to enhance performance and from interviews with experts.

classifying judgment needed to distinguish these products would have been much too arbitrary. Table A4 in the appendix assigns each product to a product group and Table A2 summarise the correspondence between product group and car's characteristics.

### 3.3.2 Data Analysis

In 2003, 1513 companies sold 412 products. In order to develop a full understanding of the variety in the niche market, the following analysis tries to answer two questions: first, it aims to describe the general supply of variety at industry's level by looking at which group provides the most variety. Moreover, since the product groups were linked with car characteristics, it analyses the generation of the variety in the different six categories of characteristics. Data on sales are not available and, therefore, both the number of products and the number of firms are used as proxy for the degree of variety of a product group or a characteristic category. The number of firms is not a trivial measure for the degree of variety because in this niche market products are distinctive from firm to firm (SEMA 2002).

Secondly, this analysis tries to survey the variety at firm level. In particular, it aims to see whether on average firms have broad ranges of products or, whether they are rather specialised. Moreover, patterns in the diversification of the firms are also explored. This last analysis provides evidence on the organization of variety within the niche market. For instance, there may be economies of scale and scope in the diversification or, on the contrary, the boundaries among different activities may be well defined. Table 3.8 shows the distribution of products across firms.

It is clear that there is an uneven distribution of number of products per category. In particular, aesthetic accessories for exterior and interior account for almost 30 % of the population. Exhaust system products, engine, electrical products for security and component vehicles do not offer a broad choice to the consumer. The "component vehicles", however, deserves a few comments. The three firms falling under this category are specialized in manufacturing whole vehicles such as replica cars. Although they purchase only one product, this could be a car or part of it entirely craft assembled and completely customised. Therefore the small number of products in this case cannot be considered as a proxy for limited variety supplied to the costumer. On the contrary, the three firms in this category are precisely the ones that generated the most diverse products.

Not surprisingly, Table 3.9 shows a similar pattern for the distribution of firms across the 20 categories. The number of firms is more evenly distributed, as the lower coefficient of variation shows, because firm sell products in more the one category. However, this table confirms that in groups like aesthetical accessories and chassis suspension and brake product there is a higher concentration of variety. The categories at the last places of the ranking are again "electrical products: security", "tire", "engine", and "component vehicles".

The second point in the agenda is to make this analysis commensurable with the previous about Ford's variety generation. As explained before, this niche market gives the opportunity to the customer to customise her car. The product car has been analysed in the previous paragraphs as bundle of characteristics grouped in eight categories (seven plus price). Following this line it could be argued that customers do not customised the car, but rather its characteristics. For this reason, I will now analyse the variety considering the product's group organised in the same characteristics categories, as put forward in the data introduction.

Table 3.10 shows the number of products that modify each single characteristic. Accessories for exterior and chassis, suspension and brakes push the characteristics "exterior" and "specification" to the first places in the ranking. "Safety" and "comfort", as the Table 3.8 showed are a very narrow group of products. Always Table 3.10 shows the number of firms selling products that change each characteristic. Results are similar besides small changes in ranking. In comparison with the other column, there is more variety in "specification" then in "performance" and more in "safety" then in "comfort-entertainment". However, the sharp dichotomy between the top three and the bottom two still holds.

In sum, from this first analysis a pattern clearly emerges: variety both in term of number of products and number of firms is concentrated in three categories

**Table 3.9** number of firm per category

| Product category | Number of firm |
|---|---|
| Accessories for exterior | 817 |
| Accessories for interior | 460 |
| Chassis, suspension and brake product | 401 |
| Engine product (external) | 350 |
| Intake and fuel system product | 348 |
| Exhaust system product | 329 |
| Wheel | 294 |
| Electrical product: lightening | 289 |
| Electrical product for engine | 276 |
| Drive train product | 271 |
| Product for engine (internal) | 227 |
| Accessories for wheels | 220 |
| Safety equipment | 216 |
| Temperature control product | 213 |
| Mobile electronic | 190 |
| Accessories for tires | 168 |
| Electrical product: security | 125 |
| Tire | 113 |
| Engine (entire or blocks) | 95 |
| Component vehicles | 3 |
| Average | 270.25 |
| Standard deviation | 164.97 |
| Coefficient of variation | 0.61 |

*Source* SEMA. My elaboration

**Table 3.10** Number of firms and products per characteristic

| Characteristic | Number of Firm | Number of products |
|---|---|---|
| Exterior | 936 | 145 |
| Performance | 654 | 105 |
| Specification | 730 | 84 |
| Interior | 460 | 30 |
| Comfort entertainment | 190 | 25 |
| Safety | 270 | 23 |

*Source* SEMA. My elaboration

(exterior, performance and specification): 81 % of the products are concentrated in these category and 61 % of the firms sell at least one product in the category accessories for exterior. A consumer is less likely to find variety to improve her safety and her comfort and entertainment.

As final comment, it is worth looking at the role of modularity. Table A4 in the appendix lists the products in the market and summarises the categories to which they were assigned. A specific technical knowledge is not necessary to understand that almost all products can be built, used, and added to a car or can substitute specific parts without requiring a modification of the car. For instance, it is possible to buy a suspension system as a new Hotchkis' Total Vehicle System already adapted to install in an old'67 Firebird. Almost all of products can be used on vehicles with no or little adaptation and without modifying any previous part of the car because they work as modules: for instance, among them, tire pressures monitoring systems, thermostats, radiators, safety belts, interior lights, MP3 players, litter baskets, wiper blades. Only the category "component vehicles" offers products that in many cases require a modification of the body of the car.

### 3.3.3 The Variety Generation at Firm Level

In the following paragraphs, I investigate the variety generation at firm level. I want to see whether companies, on average, produce many products and whether they are specialised in few categories or not. This exercise provides information on the organisation of variety within the niche market.

On average a firm in the specialty component niche market produces 22.8 products but the distributions is skewed. The top 50 companies, in term of number of products, produced on average 218.62 products, the last two hundred only one.

This structure impinges on the issue of specialisation. It is clear that few companies selling one or 200 products are active in all product groups and firms producing one or two products are obviously specialised.

The data confirm this deduction: on the one hand there is a small bunch of firms active in almost all sectors; on the other a plethora of companies that are very specialized selling only in one or two categories. Only two (0.1 %) companies sell in all 20 products' groups and 52 (3.4 %) are present in more then 14 product's groups, 924 (61 %) are selling in one or two product's groups.

It is possible to expand this analysis by investigating the rationale behind firm's diversification patterns. This is done by using a diversification matrix (Table 3.11) where the first column and the firs raw represent the specific category of products. Each figure in the matrix represents the percentage of firms, which sell the specific category of product depicted in the columns, that also sell the specific category of product highlighted in the row. These percentages are calculated as the pair wise conditional probability, measured as relative frequency that a firm is selling in two specific sub-markets. For instance, the 0.85 in italic means that 85 % of the firms selling products to modify exterior characteristics of the car are also active in the market's segment for exterior characteristics.

The first thing to note is that if a firm is selling in the "safety" or "comfort" segments, is very likely to be active in other market. These two sub-markets are the smallest in term of number products sold; therefore, firms should probably diversify in other activities to be profitable. Secondly, a firm is very likely to diversify in "exterior" characteristics and not in "safety" or "comfort". This simply depends on the critical mass in terms of firms in the exterior characteristic and to the scarcity of activity in the "safety" or "comfort". As final observation, the matrix highlights a correlation in the presence of firms in specific couples of characteristics, defined binomial group. This correlation occurs when both characteristics of each binomial group show, in comparison with the average, a high percentage value of firms diversifying in the other. Particularly noticeable is that the percentage of 85 % of firms active in "interior" sub market that sells at least one product in "exterior" as well and a reciprocal 42 % (compared with an average of 32 %) of firms active in "exterior" selling also in "interior". A second binomial group that emerge is the one between "specification" and "performance".

These two binomial groups, which are also the ones with the most variety, recall the origin of the niche market. As explained in the first part of the book, when this market emerged in the 1950s, there were two segments: the "customisers" and the "hot-rodders". The former engaged in the modification of "exterior" and "interior" mainly for aesthetic reasons, the latter modified internal parts to increase the performance of the automobile. These two groups are still discernible in the data, suggesting that today the industry is moving along the

**Table 3.11** The diversification matrix

| | Exterior | Interior | Specification | Performance | Safety | Comfort | Average |
|---|---|---|---|---|---|---|---|
| Exterior | | 0.42 | 0.42 | 0.38 | 0.26 | 0.15 | 0.32 |
| Interior | 0.85 | | 0.53 | 0.51 | 0.40 | 0.28 | 0.60 |
| Specification | 0.54 | 0.33 | | 0.62 | 0.26 | 0.17 | 0.38 |
| Performance | 0.54 | 0.36 | 0.69 | | 0.28 | 0.17 | 0.41 |
| Safety | 0.86 | 0.67 | 0.70 | 0.67 | | 0.41 | 0.66 |
| Comfort | 0.75 | 0.68 | 0.64 | 0.57 | 0.60 | | 0.65 |

*Source* SEMA. My elaboration

market trajectories dictated from the first users. Two observations should be done. First, probably descendants in consumer's habits of the first niche markets still have a similar behaviour. Therefore, there could be economies of scope in the market knowledge by selling in both segments of a binomial group.

Secondly, it is not trivial that the niche market developed along these two trajectories and this evidence deserves few observations. Customisers and hot-rodders are paradigmatic for a typical culture "on the road", freedom and emancipation in the post war US.[6] History could have gone different paths: were consumers more sensible to the threat of pollution, the niche market would probably sell today more environment-friendly devices; or maybe more security devices, if security issues would have been let to market forces instead of being subjected to legislation. Niche market could have produced components to adapt vehicles for disables, if there was a different sensitivity for people with physical impairments that can drive only special modified cars. This is not only a counterfactual "what if" analysis, because as shown in the first chapter, the automobile industry did observe at least a different niche market for specialty components emerging and disappearing, as the case of the one to convert cars into tractors.

In sum, this analysis came up with many interesting points.

1. The variety in the niche markets both in terms of the number of products and firms is concentrated around a few product groups. These products are those sold to modify exterior characteristics, performance characteristics and car specification.
2. Most of the products, above all those linked with product's groups with high variety are modules that can be easily added to a car or substitute an old part.
3. There is a dual structure of the industry: on the one hand, a bunch of firms sells many products in different market segments and on the other a plethora of small firms selling few products. The large firms are mainly active by selling product's to modify exterior performance and specification's characteristics; many firms that sell in more than one market's segment tend to be active in this group.
4. Firms selling products to modify exterior are very likely to be in the market for products for exterior and vice versa. A similar overlapping in the market occurs for firms selling product for enhance performance and modify specifications of the vehicles. This is probably due to economies of scope linked with market knowledge.

---

[6] the importance of cars for Dean Moriarty in Keruac's *On the road* is paradigmatic, or the *topos* of the car modification and car competition in many movies from and about the'1950s: for instance Grease (1978) describes the modification of a 1946 Fat Fender Ford Coupe by John Travolta's band T-birds just before a race in the 1950s.

## 3.4 Comparison

The last step of the niche market analysis, id est the comparison between the two empirical investigations, allows to provide fresh empirical evidence on the nature and operation of niche market and, secondly, on the limits in variety generation.

### *3.4.1 The Nature and Operation of Niche Market*

The first result that clearly emerged from the comparison is the evidence of a complementary variety generation. In 2003, Ford produced variety mainly for the enhancing of comfort, entertainment and safety. The niche market, made available only few products to extend these characteristics. On the contrary, the niche market provided the consumer with the possibility of modifying exterior characteristics and car's specification and performance. Customisation opportunities for these characteristics at Ford were very scarce. From the data emerges clearly a sharp division of labour in generating variety.

In order to point out an explanation I borrow from Randall and Ulrich (2001) the concepts of production-dominant variety and mediation-dominant variety. Randall and Ulrich analysed the product variety in the US Bicycle Industry. They introduce the distinction between two types of variety: production-dominant variety and mediation-dominant variety. The first is the variety that increases production costs (Stalk and Hout 1990; Cooper 1990), the second increases market mediation cost (Fisher and Ittner 1999). Therefore, in the former case economies of scale are there competitive advantage because they allow sinking the high investment in process technology. In the latter, the competitive advantage lays in the market knowledge.

As noted in the analysis of the niche market, it is reasonable to assume the existence of distinctive competencies in market knowledge for firms producing and selling products for the appearance and performance of vehicles. On the contrary, for safety and entertainment products, knowledge of the market appears not to be a competitive advantage. Probably safety characteristics are driven more by regulation rather than by consumer's preferences; entertainment and comfort, on the other hand (besides few items like a radio and a cd player), belongs to a category of products linked with new emerging technology. The market and, thus consumers' preference are not mature yet: products like GPS, Electronic parking sensors, radar detectors are still in their infant phase. Firms in the niche market therefore have not developed any market advantage. Another hypothesis could be that "safety" and "entertainment", always because are linked with new ICT technologies, required high investments in research and development that firms in the niche market cannot afford. Large companies mainly generate variety in these fields.

Large companies, on the contrary, are almost the exclusive producers of production-dominant variety. The production-dominant variety refers to the characteristics defined as "platform characteristics" in the description of variety at Ford. Variety of this type requires huge investments and should be set ex ante by the firm through the market segmentation. In the niche market there is the noticeable exception of the three firms in the product group "component vehicles" that provide variety also for platform characteristics. However, this variety produced by the components vehicle firms does not configure as production dominant. Since they craft produced cars or parts perfectly tailored to the customers their competitive advantages lay once again in the market knowledge.

A second result about the operation and nature of niche market explains that niche markets developed along two trajectories, aesthetics modification and performance enhancing. It was argued that niche market could have developed along other trajectories, like for instance production of anti-pollution devices or vehicle adaptation for people with physical impairments. It was also observed that in the history other trajectories emerged and disappeared, like for instance the production of specialty components to convert cars into tractors. This evidence suggests a particular property on the nature of the niche market. Niche markets, as both ecological theory and evolutionary approach observed, can arise only if there is a degree of heterogeneity among consumers, but, this book can add, this heterogeneity should not be too high. If preference are to diverse the niche market cannot survive: either there is a bifurcation, that is the niche market could developed its own industry if it is large enough, or it disappears. If it is not large enough can also be absorbed by another industry, as it happened in the case of specialty component niche market for tractors.

However, as final comment on the nature and organization of niche market, it should be noticed, that niche market is mainly active in this two categories (aesthetic modification and performance modification), but not only: the automobile industry has a fractal structure in the organisation of variety generation. As a geometric pattern repeated at a smaller scale, also the niche market exhibits a dual structure. Within the specialist niche market, there are few big generalist companies selling in all market segment and many small firms, which handle with few products (like specialty components vehicle): there are are new niches embedded in the niche.

### 3.4.2 The Limits of Variety Generation

Both in the description of the historical evolution of the industry and in the theoretical part, it was suggested that modularity can play a key role in the generation of variety. Evidence form the analysis supports this hypothesis. Both for Ford and in the niche market the most powerful source of variety and customisation is the use of modularity. The niche market is producing variety almost only through modular products and the large companies score better results in variety generation when they allow the consumer to customise her car with optional features. The fact, however, that the car is far to be fully modular in its design, implies that the variety introduced in this way is still marginal.

The fact that variety is concentrated mainly on modules implies that there are indeed limits of variety generation that depends on the structure of the product car. These limits were higher before the introduction of modularity. In the words of Clark the introduction of modularity has been an architectural innovation.

> The essence of an architectural innovation is the reconfiguration of an established system to link together existing components in a new way. This does not mean that the components themselves are untouched by architectural innovation. Architectural innovation is often triggered by a change in a component–perhaps size or some other subsidiary parameter of its design–that creates new interactions and new linkages with other components in the established product. The important point is that the core design concept behind each component–and the associated scientific and engineering knowledge–remain the same (Henderson and Clark 1990 p.11).

In this case, the dimension "variety" has triggered the architectural innovation. However, modularity should not be considered as a normal architectural innovation. From the data it emerged that modularity has specific features: if an architectural innovation is triggered by any dimension, e.g. size, the new structure will allow the product to growth in size, but it could lead the system to a different lock-in along other dimensions. On the contrary, in our case, the dimension triggering the changes was "variety", that is the ability of the system to change. It leads, therefore, to a structure less hierarchal and open to change along many other dimensions.

The possibility of combining many parts, by making the design more flexible, shifts the idea of variety towards the Lancastrian idea. Now the improvements in a single component already captured by the modularity in terms of variety or improvement no longer feel the constraints of the product as a whole as in the PLC story. They depend more on market conditions. Nevertheless, constrains of variety linked with the structure do not disappear. They exist in the platform, which is still the largest block in the vehicle and, as seen, shows the smaller degree of variety. Secondly, limits of the structure still exist in the evolution of the single modules, which can still an integral design.

It could be therefore argued that mainly the hierarchical structure explains the limits of variety generation. However, modularity is increasingly releasing the artefact form this boundaries. On the contrary, market forces, demand and cost of productions, are gaining importance in determining the rate and direction of variety generation. Both theoretical approaches are thus correct, but neither of them alone is complete.

# Appendix

**Table A.1** Ford's product and market segmentation

| Model | Number of trim | Category |
|---|---|---|
| Crown victoria | 4 | Passenger |
| Econoline | 18 | Mini vans and vans |
| Escape | 13 | Sport utility vehicle |
| Excursion | 31 | Sport utility vehicle |
| Expedition | 14 | Sport utility vehicle |
| Explorer | 34 | Pick-up trucks |
| Explorer sport | 16 | Sport utility vehicle |
| F-150 | 43 | Pick-up trucks |
| Focus | 16 | Passenger |
| Mustang | 14 | Passenger-luxus |
| Ranger | 25 | Pick-up trucks |
| Super duty F-250 | 33 | Pick-up trucks |
| Super duty F-350 DRW | 24 | Pick-up trucks |
| Super duty F-350 SRW | 32 | Pick-up trucks |
| Taurus—4dr Sdn LX | 1 | Passenger |
| Taurus—4dr Sdn SE | 12 | Passenger |
| Thunderbird | 5 | Passenger-luxus |
| Windstar | 6 | Passenger |
| ZX2 | 3 | Passenger |

**Table A.2** Grouping niche market products in car's characteristic

| Characteristic | Product's group |
|---|---|
| Exterior | Accessories for exterior, accessories for tyres, tyre, electrical product: lightening, wheel accessories |
| Interior | Accessories for interior |
| Performance | Chassis suspension and brakes, intake and fuel system, product for engine (external), product for temperature control |
| Specification | Special components, engine, wheel, electrical product for engine, drive train product, product for engine (internal), exhaust system products |
| Safety | Electrical product: safety, security equipment |
| comfort-entertainment | Mobile electronic |

*Source* SEMA and NADA websites. My elaboration

M. Guerzoni, *Product Variety in Automotive Industry*,
SpringerBriefs in Business, DOI: 10.1007/978-3-319-01907-9,

**Table A.3** Variables in the dataset

| Variable | Type of variable | Notes on the variable | Variable | Type of variable | Notes on the variable |
|---|---|---|---|---|---|
| *Price* | | | *Entertainment* | | |
| Base MSRP price | Continuous | In the analysis | AM/FM Stereo | | In the analysis |
| Base invoice price | Continuous | Replication | Cassette Player | | In the analysis |
| Destination charge | Continuous | No interest | CD Player | | In the analysis |
| | | | CD Changer | | In the analysis |
| *Specifications* | | | | | |
| Engine type | Multinomial | In the analysis | *Performance* | | |
| Displacement | Multinomial | In the analysis | Horsepower | Continuous | In the analysis |
| Fuel system | Multinomial | In the analysis | Torque | Continuous | In the analysis |
| Drive train e | | In the analysis | Dead weight hitch—max trailer | Continuous | In the analysis |
| Transmission type | Multinomial | In the analysis | Dead weight hitch—max tongue | Continuous | In the analysis |
| Transmission description | | In the analysis | Distribution hitch—max trailer | Continuous | Tautological |
| EPA fuel economy est.—city | Continuous | Missing values | Distribution hitch—max tongue | Continuous | Tautological |
| EPA fuel economy est.—hwy | Continuous | Missing values | | | |
| Fuel tank capacity | Continuous | In the analysis | *Interior* | | |
| Cruising range—city | Continuous | Missing values | Passenger capacity | Discrete | In the analysis |
| Cruising range—hwy | Continuous | Missing values | Cargo volume | Continuous | In the analysis |
| Fuel type | Multinomial | No differentiation | Front head room | Continuous | In the analysis |
| | | | Second row head room | Continuous | In the analysis |
| *Safety* | | | Front leg room | Continuous | In the analysis |
| Air bag—driver—front | Binary | No differentiation | Second row leg room | Continuous | In the analysis |
| Air bag—passenger—front | Binary | No differentiation | Front hip room | Continuous | In the analysis |
| Air bag—passenger switch | Binary | In the analysis | Second row hip room | Continuous | In the analysis |
| Air bag—side—body—front | Binary | Missing values | Front shoulder room . | Continuous | In the analysis |
| Air bag—side—head only | Binary | Missing values | Second row shoulder room | Continuous | In the analysis |
| ABS brakes—4 wheel | Binary | Missing value | | | |

(continued)

Table A.3 (continued)

| Variable | Type of variable | Notes on the variable | Variable | Type of variable | Notes on the variable |
|---|---|---|---|---|---|
| ABS brakes—rear wheels only | Binary | In the analysis | *Exterior* | | |
| Traction control | | In the analysis | Wheelbase | Continuous | In the analysis |
| Child Safety locks | Binary | In the analysis | Length | Continuous | In the analysis |
| Headlights—automatically on | Binary | No differentiation | Width, max w/o mirrors | Continuous | In the analysis |
| Headlights—daytime running | Binary | No differentiation | Height—overall 54.5 | Continuous | In the analysis |
| lights | | | Base curb weight 3,920 | Continuous | Missing values |
| *Comfort* | | | Ground clearance | Continuous | Missing values |
| Cruise control | | In the analysis | Turning diameter—curb to curb 36.1 | Continuous | Missing values |
| Air conditioning—front | Binary | In the analysis | Front tires | Multinomial | In the analysis |
| Automatic climate control | Binary | In the analysis | Rear tires | Multinomial | Same information |
| Dual zone climate control | Binary | Never available | Sunroof / moonroof | Binary | In the analysis |
| Remote keyless entry | Binary | In the analysis | Rear window defroster | Binary | In the analysis |
| Vehicle anti-theft protection | Binary | In the analysis | | | |
| Power windows | Binary | Standard | *Warranty* | | |
| Power door locks | Binary | Standard | Basic 4 years | Multinomial | Not relevant |
| Power steering | Binary | Standard | Drivetrain 4 years | Multinomial | Not relevant |
| Power mirrors—driver | Binary | Standard | Corrosion 5 years | Multinomial | Not relevant |
| Power mirrors—passenger | Binary | Standard | Roadside assitance | Multinomial | Not relevant |
| Power seats—driver | Binary | In the analysis | | | |
| Power seats—passenger | Binary | Tautological | | | |
| Leather seats | Binary | In the analysis | | | |
| Steering wheel audio controls | Binary | Never available | | | |
| Navigation system | Binary | Never available | | | |
| Universal garage door opener | | In the analysis | | | |

*Source* NADA website. My elaboration

**Table A.4** Products sold in the niche market and categorisation. Source SEMA website. My elaboration

| Product | Category | Product | Category |
|---|---|---|---|
| Air_dams | Accessories for exterior | Electronic parking sensors | Mobile electronic |
| Bed drawers | Accessories for exterior | Equalizers | Mobile electronic |
| Bed extenders | Accessories for exterior | Gaskets/washers | Mobile electronic |
| Bed rail caps | Accessories for exterior | Head Units/receivers | Mobile electronic |
| Bed_mats | Accessories for exterior | Insulating material | Mobile electronic |
| Bedliners–drop-In/carpet | Accessories for exterior | Miscellaneous hardware | Mobile electronic |
| Bedlinersspray-on | Accessories for exterior | Monitors | Mobile electronic |
| Body decals/graphics | Accessories for exterior | MP3 players | Mobile electronic |
| Body parts/panels–fiberglass | Accessories for exterior | Navigation systems/GPS | Mobile electronic |
| Body parts/panels–sheet metal | Accessories for exterior | Radar detectors | Mobile electronic |
| Body styling kits | Accessories for exterior | Radar scramblers | Mobile electronic |
| Body trim/molding | Accessories for exterior | Speakers | Mobile electronic |
| Bras/masks | Accessories for exterior | Subwoofers | Mobile electronic |
| Bug deflectors/bug guards | Accessories for exterior | Telematics | Mobile electronic |
| Bumper covers | Accessories for exterior | TV tuners | Mobile electronic |
| Bumpers | Accessories for exterior | Video antennas | Mobile electronic |
| Cab extenders | Accessories for exterior | Videocassette players | Mobile electronic |
| Camper shells | Accessories for exterior | Alternators | Electrical product for engine |
| Car/truck covers | Accessories for exterior | Batteries | Electrical product for engine |
| Cargo/trunk organizers | Accessories for exterior | Computer chipsvehicle | Electrical product for engine |
| Conversions | Accessories for exterior | Computersvehicle | Electrical product for engine |
| Convertible tops | Accessories for exterior | Cruise control units | Electrical product for engine |
| Custom truck beds | Accessories for exterior | Electronic control modules | Electrical product for engine |
| Doors | Accessories for exterior | Generators | Electrical product for engine |
| Fascias | Accessories for exterior | Insulating productselectronic | Electrical product for engine |
| Fender trim and molding | Accessories for exterior | Sensors | Electrical product for engine |
| Flaresfender/body/etc. | Accessories for exterior | Solenoids | Electrical product for engine |

(continued)

**Table A.4** (continued)

| Product | Category | Product | Category |
|---|---|---|---|
| Fuel filler doors | Accessories for exterior | Starters | Electrical product for engine |
| Gas caps | Accessories for exterior | Voltage regulators | Electrical product for engine |
| Gasketsbody accessories | Accessories for exterior | Wiring harnesses | Electrical product for engine |
| Gold trim | Accessories for exterior | Automotive bulbs | Electrical product: lightening |
| Grille guards/brush guards | Accessories for exterior | Backup lights | Electrical product: lightening |
| Grille guards/push bars/brush guards/bull bars | Accessories for exterior | Center/high-mounted stoplights | Electrical product: lightening |
| Grilles | Accessories for exterior | Driving/foglights | Electrical product: lightening |
| Ground effects | Accessories for exterior | Halogen bulbs | Electrical product: lightening |
| Hardtops | Accessories for exterior | Handheld lights | Electrical product: lightening |
| Headache racks | Accessories for exterior | HID/Xenon headlights | Electrical product: lightening |
| Headlight lenses | Accessories for exterior | Industrial/work lights | Electrical product: lightening |
| Hitch covers | Accessories for exterior | Interior lighting | Electrical product: lightening |
| Hoods | Accessories for exterior | leds | Electrical product: lightening |
| Ladder racks | Accessories for exterior | Lighting harnesses | Electrical product: lightening |
| License plate frames | Accessories for exterior | Neon lighting | Electrical product: lightening |
| License plates | Accessories for exterior | Off-road lights | Electrical product: lightening |
| Lift gates | Accessories for exterior | Projector headlights | Electrical product: lightening |
| Light bars | Accessories for exterior | Rear foglights | Electrical product: lightening |
| Loading ramps | Accessories for exterior | Sealed beam bulbs | Electrical product: lightening |
| Louvers | Accessories for exterior | Strip lighting | Electrical product: lightening |
| Luggage racks | Accessories for exterior | Taillights | Electrical product: lightening |
| Mirros | Accessories for exterior | Power window/door/trunk locks | Electrical product: safety |
| Mud flaps/splash guards | Accessories for exterior | Security devices/alarms | Electrical product: safety |
| Nerf bars, accent bars | Accessories for exterior | Diesel engines | Engine |
| Pinstriping | Accessories for exterior | Engine blocks | Engine |
| Racks/carriers | Accessories for exterior | Enginescomplete | Engine |
| Rear decorating lenses | Accessories for exterior | Air cleaners/accessories | Product for engine (external) |

(continued)

**Table A.4** (continued)

| Product | Category | Product | Category |
|---|---|---|---|
| Rocker panel trim | Accessories for exterior | Air conditioning compressors | Product for engine (external) |
| Running boards | Accessories for exterior | Air conditioning hoses | Product for engine (external) |
| Simulated convertibles/landaus | Accessories for exterior | Air conditioning units | Product for engine (external) |
| Spare tire covers | Accessories for exterior | Dipsticks | Product for engine (external) |
| Spoilers | Accessories for exterior | Engine adapters | Product for engine (external) |
| Stepscab/side/tire | Accessories for exterior | Engine dress-up items | Product for engine (external) |
| Sunroof deflectors | Accessories for exterior | Engine hoses | Product for engine (external) |
| Sunroofselectric power sliding | Accessories for exterior | Harmonic balancers/dampers | Product for engine (external) |
| Sunroofspop-up tilting | Accessories for exterior | Hose clamps | Product for engine (external) |
| Sunroofs spoiler top-mount | Accessories for exterior | Motor mounts | Product for engine (external) |
| SunroofsT-top | Accessories for exterior | Oil caps | Product for engine (external) |
| Tailgate nets | Accessories for exterior | Oil pans | Product for engine (external) |
| Tailgate protectors | Accessories for exterior | Timing covers | Product for engine (external) |
| Tailgates | Accessories for exterior | Valve covers | Product for engine (external) |
| Taillight covers | Accessories for exterior | Boltsrod, head, etc. | Product for engine (internal) |
| Taillight lenses | Accessories for exterior | Camshafts | Product for engine (internal) |
| Tape friction/masking/striping | Accessories for exterior | Connecting rods | Product for engine (internal) |
| Tie-downs | Accessories for exterior | Crankshafts | Product for engine (internal) |
| Tire covers | Accessories for exterior | Cylinder heads | Product for engine (internal) |
| Tonneau covers | Accessories for exterior | Engine bearings | Product for engine (internal) |
| Toolboxes/bed boxes | Accessories for exterior | Engine rebuilding kits | Product for engine (internal) |
| TopsT-tops/bikini tops/umbrella tops | Accessories for exterior | Freeze plugs | Product for engine (internal) |
| Towing accessories | Accessories for exterior | Gasketsinternal engine | Product for engine (internal) |
| Truck bed accessories | Accessories for exterior | Gear drives | Product for engine (internal) |
| Truck bed bars | Accessories for exterior | Lifters | Product for engine (internal) |
| Truck caps | Accessories for exterior | Main caps | Product for engine (internal) |
| Truck slider windows | Accessories for exterior | Oil pumps | Product for engine (internal) |

(continued)

**Table A.4** (continued)

| Product | Category | Product | Category |
|---|---|---|---|
| Vents | Accessories for exterior | Piston ring sets | Product for engine (internal) |
| Visors | Accessories for exterior | Pistons | Product for engine (internal) |
| Winches | Accessories for exterior | Push rods | Product for engine (internal) |
| Window boots | Accessories for exterior | Rocker arms | Product for engine (internal) |
| Window deflectors | Accessories for exterior | Timing chains/belts/gears | Product for engine (internal) |
| Window tint film | Accessories for exterior | Valve springs | Product for engine (internal) |
| Windows | Accessories for exterior | Valves | Product for engine (internal) |
| Wings | Accessories for exterior | Windage trays/baffles | Product for engine (internal) |
| Wiper blades | Accessories for exterior | Catalytic converters | Exhaust system product |
| 12-Volt accessories | Accessories for interior | Engine/exhaust brakes | Exhaust system product |
| Armrest | Accessories for interior | Exhaust headers | Exhaust system product |
| Car/truck floor mats | Accessories for interior | Exhaust manifolds | Exhaust system product |
| Cargo area liners | Accessories for interior | Exhaust pipes/tubing | Exhaust system product |
| Carpeting | Accessories for interior | Exhaust tips | Exhaust system product |
| Consoles | Accessories for interior | Gasketsexhaust system | Exhaust system product |
| Cupholders | Accessories for interior | Hangers/clamps | Exhaust system product |
| Dashboard covers | Accessories for interior | Mufflers | Exhaust system product |
| Dashboard trim/overlays | Accessories for interior | Air snorkels | Intake and fuel system |
| Door panels | Accessories for interior | Auxiliary tanks | Intake and fuel system |
| Doorsill protectors | Accessories for interior | Boost controls | Intake and fuel system |
| Headliners | Accessories for interior | Carburetor adapters | Intake and fuel system |
| Interior graphics | Accessories for interior | Carburetor baseplates | Intake and fuel system |
| Leather | Accessories for interior | Carburetors | Intake and fuel system |
| Litter baskets | Accessories for interior | Cold air intake systems | Intake and fuel system |
| Seat covers | Accessories for interior | Electronic control unitsfuel system | Intake and fuel system |
| Seat heaters | Accessories for interior | Fuel injection systems | Intake and fuel system |
| Seat upholstery | Accessories for interior | Fuel injectors | Intake and fuel system |

(continued)

**Table A.4** (continued)

| Product | Category | Product | Category |
|---|---|---|---|
| Seatscustom | Accessories for interior | Fuel lines | Intake and fuel system |
| Steering wheel covers | Accessories for interior | Fuel pressure regulators | Intake and fuel system |
| Steering wheels | Accessories for interior | Fuel pumps | Intake and fuel system |
| Storage boxes/compartments | Accessories for interior | Fuel tanks | Intake and fuel system |
| Tables/snack trays | Accessories for interior | Gasketsfuel system | Intake and fuel system |
| Trunk liners/mats | Accessories for interior | Intake manifolds | Intake and fuel system |
| Trunk organizers | Accessories for interior | Intercoolers | Intake and fuel system |
| Van furniture | Accessories for interior | Nitrous oxide systems | Intake and fuel system |
| Weatherstripping | Accessories for interior | Octane boosters | Intake and fuel system |
| WIndow | Accessories for interior | Racing fuels | Intake and fuel system |
| Window blinds/curtains | Accessories for interior | Superchargers | Intake and fuel system |
| Window shades | Accessories for interior | Throttle bodies | Intake and fuel system |
| Air bags suspension | Chassis suspension and brakes | Turbochargers | Intake and fuel system |
| Air brake parts | Chassis suspension and brakes | Velocity stacks | Intake and fuel system |
| Antilock brake systems | Chassis suspension and brakes | Airbags | Safety |
| Bearings wheel/axle/etc. | Chassis suspension and brakes | Alcohol detectors | Safety |
| Brake cables | Chassis suspension and brakes | Backup cameras | Safety |
| Brake calipers | Chassis suspension and brakes | Blankets | Safety |
| Brake conversion kits | Chassis suspension and brakes | Camera vision products | Safety |
| Brake cylinders | Chassis suspension and brakes | Child restraints | Safety |
| Brake drums | Chassis suspension and brakes | Child seats | Safety |
| Brake lines/clamps | Chassis suspension and brakes | Crash test dummies | Safety |
| Brake parts, disc | Chassis suspension and brakes | Fire extinguishers | Safety |
| Brake rotors | Chassis suspension and brakes | Fresh air breathing apparatus | Safety |
| Brake shoes/pads | Chassis suspension and brakes | Heat shields/firewall insulation | Safety |
| Bushings/rubber | Chassis suspension and brakes | Helmets | Safety |
| Bushings/synthetic/poly | Chassis suspension and brakes | Hood locks | Safety |

(continued)

**Table A.4** (continued)

| Product | Category | Product | Category |
|---|---|---|---|
| Chassis kits | Chassis suspension and brakes | Parachutes | Safety |
| Coil springs | Chassis suspension and brakes | Pet restraints | Safety |
| Computer-controlled ride control systems | Chassis suspension and brakes | Racing seats | Safety |
| Control arms | Chassis suspension and brakes | Racing shoes/apparel | Safety |
| CV boots | Chassis suspension and brakes | Roll bars | Safety |
| CV joints | Chassis suspension and brakes | Roll cages | Safety |
| Frames vehicle | Chassis suspension and brakes | Safety belts/harnesses | Safety |
| Idler arm gussets | Chassis suspension and brakes | Scatter shields/guards/blankets | Safety |
| Idler arms | Chassis suspension and brakes | Replica cars | Specialty component |
| Independent front suspension | Chassis suspension and brakes | Condensers | Product for temperature control |
| Independent rear suspension | Chassis suspension and brakes | Conventional fans | Product for temperature control |
| Leaf springs | Chassis suspension and brakes | Defoggers/defrosters | Product for temperature control |
| Lift kits | Chassis suspension and brakes | Electric fans | Product for temperature control |
| Line lock system | Chassis suspension and brakes | Engine cooling systems | Product for temperature control |
| Lowering kits | Chassis suspension and brakes | Fan belts | Product for temperature control |
| Multiple shock kits | Chassis suspension and brakes | Gasketscooling system | Product for temperature control |
| Off-road suspension kits | Chassis suspension and brakes | Hosescooling system | Product for temperature control |
| Race car chassis building supplies | Chassis suspension and brakes | Oil coolers | Product for temperature control |
| Rubber springs and suspension | Chassis suspension and brakes | Pulleys | Product for temperature control |
| Shackles | Chassis suspension and brakes | Radiator caps | Product for temperature control |
| Shock absorbers/struts | Chassis suspension and brakes | Radiators | Product for temperature control |
| Skid plates | Chassis suspension and brakes | Thermostats | Product for temperature control |
| Steering boxes | Chassis suspension and brakes | Transmission coolers | Product for temperature control |
| Steering columns | Chassis suspension and brakes | Transmissions | Product for temperature control |
| Steering conversion kits | Chassis suspension and brakes | Water pumps | Product for temperature control |
| Steering stabilizers | Chassis suspension and brakes | Agricultural tires | Tire |
| Steering U-joints/assemblies | Chassis suspension and brakes | Antique/classic/vintage tires | Tire |

(continued)

**Table A.4** (continued)

| Product | Category | Product | Category |
|---|---|---|---|
| Strut tower braces | Chassis suspension and brakes | ATV/ATC tires | Tire |
| Subframe connectors | Chassis suspension and brakes | Drag racing tires | Tire |
| Subframes | Chassis suspension and brakes | Heavy truck tires | Tire |
| Sway bars | Chassis suspension and brakes | High-performance street tires | Tire |
| Tie-rod ends | Chassis suspension and brakes | Industrial tires | Tire |
| Torsion bars | Chassis suspension and brakes | Light-truck/SUV tires | Tire |
| Traction bars/controls | Chassis suspension and brakes | Off-road tires | Tire |
| Trusses | Chassis suspension and brakes | Oval track/road racing tires | Tire |
| Axles | Drivetrain product | Passenger car tires | Tire |
| Bell housings | Drivetrain product | Sand sports tires | Tire |
| Clutches | Drivetrain product | Inner tubes | Accessories for tyres |
| Differential covers | Drivetrain product | Pressure gauges | Accessories for tyres |
| Differential gears | Drivetrain product | Snow chains | Accessories for tyres |
| Driveshaft loops | Drivetrain product | Tire covers | Accessories for tyres |
| Driveshafts/half shafts | Drivetrain product | Tire deflators | Accessories for tyres |
| Flexplates | Drivetrain product | Tire inflating systems | Accessories for tyres |
| Flywheels | Drivetrain product | Tire inflation accessories | Accessories for tyres |
| Gaskets drivetrain | Drivetrain product | Tire lubricants | Accessories for tyres |
| Hubsaxle | Drivetrain product | Tire pressure monitoring systems | Accessories for tyres |
| Linkage kits | Drivetrain product | Tire pumps | Accessories for tyres |
| Overdrive/underdrive units | Drivetrain product | Tire repair materials | Accessories for tyres |
| Positive traction devices | Drivetrain product | Tire sealants | Accessories for tyres |
| Pressure plates | drivetrain product | Tire studs | Accessories for tyres |
| Ring-and-pinion gears | Drivetrain product | Valve stem caps | Accessories for tyres |
| Shift improvement kitsperformance/manual/auto | Drivetrain product | Valve stems | Accessories for tyres |
| Shifters automatic | Drivetrain product | Aluminum wheelscast/forged | Wheel |
| Shifters manual | Drivetrain product | Composite wheels | Wheel |

(continued)

**Table A.4** (continued)

| Product | Category | Product | Category |
|---|---|---|---|
| Throwout bearings standard/hydraulic | Drivetrain product | Dual wheels | Wheel |
| Torque converters | Drivetrain product | Modular wheels | Wheel |
| Transaxles | Drivetrain product | Original (OEM)/replica wheels | Wheel |
| Transfer cases | Drivetrain product | Steel wheels | Wheel |
| Transmission components | Drivetrain product | Three-piece wheels | Wheel |
| Transmission mounts | Drivetrain product | Two-piece wheels | Wheel |
| Transmissions automatic | Drivetrain product | Wire wheels | Wheel |
| Transmissions manual | Drivetrain product | Accent lighting–wheels | Accessories for wheels |
| Universal joints | Drivetrain product | Lug nuts | Accessories for wheels |
| Amplifiers | Mobile electronic | Wheel adapters | Accessories for wheels |
| Antennas | Mobile electronic | Wheel center caps | Accessories for wheels |
| Cameras | Mobile electronic | Wheel covers | Accessories for wheels |
| Cassette players/recorders | Mobile electronic | Wheel hubs | Accessories for wheels |
| CD changers | Mobile electronic | Wheel locks | Accessories for wheels |
| CD players/recorders | Mobile electronic | Wheel lug pattern tools | Accessories for wheels |
| Crossover networks | Mobile electronic | Wheel spacers | Accessories for wheels |
| DVD players | Mobile electronic | Wheel spinners | Accessories for wheels |

# Conclusion

In this book I carried out an investigation on product variety in the U.S. automobile industry. This investigation is interesting for two reasons. The first reason is the presence in the industry of a niche market that provides opportunities for post sales customisation. This peculiar structure allows the development a new type of analysis of product variety. Secondly, niche markets are an important source of qualitative variety, according to Saviotti the most important factor for the economic growth. In order to further develop this issue, I aimed to shed fresh empirical light both on the nature of niche market in this industry and on the limits of variety generation.

Chapters 1 and 2 provided the historical context and the theoretical background to define a framework for the data analysis. Chapter 1 explained how variety broadened over the last century and how a niche market for specialty components developed. From the description of the sector emerged a paradox that is the co-existence of a growing niche market that provide post sale customisation and of the successful effort of increasing variety in the automobile industry. Chapter 2 described two different theoretical approaches to the analysis of variety that lead to different hypotheses on its nature and on the nature of its limits. PLC theories consider variety as improvement in product and suggest that its limits depend on the hierarchical structure of the product design, whilst Lancastrian approach stresses the interplay of market forces. The attempt both of finding a reason for the paradox and of testing these two different frameworks has steered the route of the empirical analysis.

In Chap. 3, I first explained why methodologies used in other studies that addressed similarly issues could not be applied to this research and, thus, why the need of a new tool emerged. I, thereafter, introduced the niche market analysis as tool of analysis and I discussed its assumption and limits.

Finally, Chap. 3 disclosed the results of the empirical analysis. Concerning the nature and the operation of the niche market, it first highlighted a sharp division of labour between large companies and the niche market: the niche market generates variety where market knowledge is a key competitive advantage. Large companies, as represented by Ford, on the contrary, produced two kinds of variety: "variety as customisation", in those fields where the economies of scale are not

M. Guerzoni, *Product Variety in Automotive Industry*,
SpringerBriefs in Business, DOI: 10.1007/978-3-319-01907-9,

crucial but the niche market did not develop any market knowledge and "variety as segmentation" where economies of scale are important and variety is set before the production. "Variety as customisation" is relatively high and "variety as segmentation" is narrow. From the analysis, it also emerged that this niche market is not a homogenous entity but, as a fractal, it presents again at a smaller scale the same dualism between large and small companies. A last result concerned the relationship between demand heterogeneity and emergence and survival of niche markets: heterogeneity is required but preference should be not too diverse.

Concerning the limits of variety generation, as premise, it was shown that, in line with the predictions of the theory, modularity plays a key role in the customisation process. The paradox put forward in the first chapter finds a solution. For large companies modularity has been a double edge sword: the same modularity introduced by car producers to broaden the variety's supply allowed the development of a niche market that can better exploit it. Modularity reduces the fixed cost and, therefore the importance of economies of scale, as competitive advantage for large companies. The fact that only by introducing modularity, id est an architectural innovation, firms can produce more variety implies that the limits of production are indeed in the structure of the product, as explained by PLC theory. However, it was explained that modularity itself makes the hierarchy more flexible and shifts the limits of variety from the design to the Lancastrian market forces. *Rebus sic stantibus*, the two limits coexist and, since the design of the car is far from being modular, the limits of design hierarchy still play a major role. In addition, even if the future car design is fully modular, PLC theory will still hold for the explanation of the evolution of variety in the single components.

These results have few implications for this industry and call for more research.

First, they warn expertises of the industry about the possible disruption coming from a further modularisation. In many industries, where modularity has reached a higher degree, there is a competitive tension between the owner of platform and producers of devices for post sales product. On the one hand post sale customisation can be considered as a way to increase the appealing of the product that generates positive externalities and, therefore, more sales for platform owner. On the other hand, it can be seen as a threat because it can shift the perceived valued (thus, the highest mark up) from the platform to the module. This goes to detriment of the profits of the platform owner. The relation between hardware and software in the computer industry is paradigmatic for this competitive tension.

This book has a poor dynamic dimension and it cannot forecast the future evolution of modularity. For the moment, modularity is limited and niche market negligible compared with the figure of the large producers. It is pointed out, however, the strategic importance of the direction of modularity. Large firms compete in variety by producing increasing modular artefact, but this results suggest that the niche market can partly appropriate of the benefits from increased modularity. Results also suggest large companies a strategy in order to have higher appropiability of benefits from variety generation gained through modularity. Further development of modularity should concern those components where the niche market does not have any market competitive advantages.[1] This issue calls for more empirical research to understand along what kinds of path the design of vehicles can evolve in the industry.

I can add two final comments, which are not industry specific. First, the new methodology applied here can be generalized to other sector, *mutatis mutandis*. The automobile industry is not the only industry that presents this dualism in its industrial structure: the bicycle industry show a similar dualism, in computer industry there is niche market for customising the machine (by adding more memories, buying different devices, etc.) and even in the furniture industry there is the possibility of buying modules for self assembling and of a concrete "do it yourself". Moreover, as seen in these conclusions, niche market analysis can be a tool for a researcher interested in the economics of variety generations, but it can also be a way to assess the strategic position of firm in the generation of variety and a promising tool for marketing research on consumer preferences.

Finally, the evidence of the coexistence of different types of limits has few implications. It implies that the distinction operated in literature makes little sense: variety as innovation, in the PLC theory, and variety as differentiation, in the Chamberlain and Lancastrian approaches, occur together and are hardly distinguishable. This book therefore calls for a holistic approach that take into account the two varieties and their interplay. This support the view of those scientists (Freeman 1994, for instance p. 480) that suggested that the ancient dichotomy demand-pull/technology-push in innovation studies should be buried. Modularity, by releasing the variety from architectural boundaries empowered the market forces and thus demand to pull variety and innovation; however the persistence of constrains erected by technological trajectory and design architecture has still to be considered. This result could be generalise to other industries since it is a challenge to scientific community to bridge the economics of variety and the economic of innovation and technical change, although they are often considered as alternative. In sum, this book supports the view of the existence of a strong link between variety and innovation that should be taken into account in economic theory.

---

[1] The recent attempt of GM of providing information to disables for adapting vehicles can be read as strategic choice in this direction (GM website). In US there are 33.7 million of persons with limitations in usual activities (*source* Centre for Decease Control and Prevention, US Department of Health and Human Services, website).

# References

## Journals, books, and articles

Agrawal, M., Kumaresh, T. V., & Mercer, G. A. (2001). The false promise of mass customization. *The McKinsey Quarterly, 3*, 62–71.

Armstrong, G., & Kotler P. (2003). *Marketing: An introduction* (6th ed.). New York: Prentice Hall.

Abernathy, W. J., & Utterback, J. M. (1975). A dynamic model of process and product innovation. *Omega, 3*(6), 639–656.

Abernathy, W. J., & Utterback, J. M. (1978). Patterns of industrial innovation. *Technology Review, 80*(7), 40–47.

Abernathy, W. J., Clark, K. B., & Kantrow, A. M. (1983). *Industrial renaissance: Producing a competitive future for America*. New York: Basic Books.

Adner, R. (2002). When are technologies disruptive: A demand-based view of the emergence of competition. *Strategic Management Journal, 23*, 667–688.

Alexander, C. (1964). *Notes on the synthesis of form*. Cambridge: Harvard University Press.

Anderson, P., & Tushman, M. L. (1990). Technological discontinuities and dominant design: A cyclical model of technological change. *Administrative Science Quarterly, 35*(4), 604–633.

Averitt, R. (1968). *The dual economy*. New York: Norton.

Baldwin, C. Y., & Clark, K. B. (1997). Managing in an age of modularity. *Harvard Business Review, 75*(5), 84–93. (September–October).

Baldwin, C. Y., & Clark, K. B. (2000). *Design rules. The power of modularity*. Cambridge: MIT Press.

Barnes, J.R. (2001). *World class manufacturing as a necessary but insufficient condition for industry success: A case study of the South African automotive components industry*. Ph.D. Thesis, School of Development Studies, University of Natal, Pietermaritzburg.

Berger, S., & Piore, M. (1980). *Dualism and discontinuity in industrial societies*. Cambridge: Cambridge University Press.

Best, M. (1990). *The new competition*. Oxford: Polity Press.

Boone, C., Carroll, G. R., & van Witteloostuijn, A. (2002). Environmental resource distributions and the market partitioning of Dutch daily newspaper organizations. *American Sociological Review, 67*, 408–431.

Burns, T., & Stalker, G. M. (1966). *The management of innovation*. London: Tavistock publications.

Calcagno, M. (2002). *Dynamics of modularity. A critical approach*. Working paper presented at the 2nd EURAM Conference, Stockholm, May 2002.

Camuffo, A. (2002). *Globalization, outsourcing and modularity in the auto industry*. Paper presented at the 2nd EURAM Conference, Stockholm, May 2002.

M. Guerzoni, *Product Variety in Automotive Industry*,
SpringerBriefs in Business, DOI: 10.1007/978-3-319-01907-9,

Camuffo, A., & Volpato, G. (2001). *The FIAT-GM strategic alliance*. Paper presented at the 9th GERPISA International Colloquium, Reconfiguring the Auto Industry: Merge and Acquisition, Aliiances and Exit, Palais du Luxembourg, Paris 7–9 June.

Carroll, G. R. (1985). Concentration and specialization: Dynamics of niche width in populations of organizations. *American Journal of Sociology, 90*, 1262–1283.

Carroll, G., & Hannah, M. T. (1995). Resource partitioning. In G. Carroll & M. T. Hannan (Eds.), *Organizations in industry. Strategy, structure, and selection* (pp. 215–221). New York: Oxford University Press.

Cebon, P., Hauptman, O., & Shekhar, C. (2001). *Industries in the making: Product modularity, technological innovation and the product life cycle*. Report No MBS 2001-04, for Academic researchers. Melbourne: Melbourne Business School.

Chamberlain, E. (1933). *The theory of monopolistic competition*. Cambridge: Harvard University Press.

Chan, K. (2001). *Customization effects of e-supply chain design with reference to the bib at the automobile industry*. Paper presented at the 9th GERPISA Internationla Colloquium, Reconfiguring the Auto Industry: Merge and Acquisition, Alliances and Exit, Palais du Luxembourg, Paris.

Christensen, C. M. (1997). *The innovator's dilemma: When new technologies cause great firms to fail*. Cambridge: Harvard Business School Press.

Clark, K. (1985). The interaction of design hierarchies and market concepts in technological evolution. *Research Policy, 14*, 235–251.

Cooper, R. (1990). Cost classification in unit-based and activity-based manufacturing cost system. *Journal of Cost Management, 6*, 6–14.

Cusumano, R. M. (1988). *Manufacturing innovation: Lessons from the Japanese auto industry*. Fall: Sloan Management Review.

d'Aspremont, C., Jaskold-Gabszewicz, J., & Thisse, J.-F. (1979). On hotelling's stability in competition. *Econometrica, 47*, 1145–1150.

Davis, S. M. (1987). *Future perfect*. Reading: Addison-Wesley.

Dichter, E. (1965). Discovery the inner Jones. *Harvard Business Review, May–June*(3), 6–10.

Dixit, A. K., & Stiglitz, J. E. (1977). Monopolistic competition and optimum product diversity. *American Economic Review, 67*, 297–308.

Dobrev, S. D. & Carroll G. R. (2000). *Size (and competition) among organizations: Modelling scale-based selection among automobile producers in four major countries, 1885–1981*. Paper presented at the Academy of Management Meetings, Toronto.

Dobrev, S. D., Carrol, G., & Kim, T. Y. (2002). Evolution of organizational niches: U.S. automobile manufacturers, 1885–1981. *Administrative Science Quarterly, 47*, 233–264.

Dobrev, S. D., Kim, T. Y., & Hannan, M. T. (2001). Dynamics of niche width and resource partitioning. *American Journal of Sociology, 106*, 1299–1337.

Dosi, G. (1982). Technological paradigms and technological trajectories: A suggested interpretation of the determinants and directions of technical change. *Research Policy, 11*(3), 147–162.

Dosi, G., Marsili, O., Orsenigo, L., & Salvatori, R. (1995). Learning, market selection, and the evolution of industrial structure. *Small Business Economics, 7*, 411–436.

Farhi, S. (1998). French suppliers has eye on U.S. *Automotive News, 73*(5812), 24.

Faulkner, P. (1980). *Title against the age: An introduction to William Morris*. London: Publisher Allen and Unwin.

Fine, C. H., Clair, R. S., Lafrance, J. C., & Hilderbrand, D. (1996). *The US automobile manifucturing industry*. Washington US Department of Commerce, Office of Technology Policy.

Fisher, M., & Ittner, C. (1999). The impact of product variety on automobile assembly operations: Analysis and evidence. *Management Science, 45*(6), 771–786.

Foray, D., & Grübler, A. (1990). Morphological analysis, diffusion and lock-out of technologies: Ferrous casting in France and the FRG. *Research Policy, 19*, 535–550.

Freeman, C. (1994). The economics of technical change. *Cambridge Journal of Economics, 18*, 479–482.

Frenken, K., & Leydesdorff, L. (2000). Scaling trajectories in civil aircraft (1913–1997). *Research Policy, 29*(3), 331–348.

Frenken, K., Saviotti, P. P., & Trommetter, M. (1999). Variety and niche creation in aircraft, helicopters, motorcycles and microcomputers. *Research Policy, 28*(5), 469–488.

Fujimoto, T., & Takeishi, A. (2001). *Modularization in the auto industry: Interlinked multiple hierarchies of product, production and suppliers system.* CIRJE-F-107 Discussion Paper, Tokyo University, Tokyo.

Galbraith, J. K. (1967). *The new industrial state*. Boston: Huntington Mifflin.

Gawer, M., & Cusumano, A. (2002). *Platform leadership: How intel, microsoft, and cisco drive industry innovation*. Boston: Harvard Business School Press.

Hannan, M. T. (1997). Inertia, density and the structure of organizational populations: Entries in European automobile industry, 1886–1981. *Organization Studies, 18*, 193–228.

Hannan, M. T., Carroll, G. R., Dundon, E., & Torres, J. C. (1995). Organizational evolution in multinational context: Entries of automobile manufacturers in Belgium, France, Germany and Italy. *American Sociological Review, 60*, 509–528.

Hart, O. D. (1985). Monopolistic competition in the spirit of Chamberlain. A general model. *Review of Economic Studies, 52*, 529–46.

Helper, S., MacDuffie J. P., Pil, F., Sako, M., Takeishi, A,. & Warburton, M. (1999). *Project report: Modularization and outsourcing: Implications for the future of automotive assembly.* Paper presented for the IMVP Annual Forum, MIT, Boston, 6–7 October 1999.

Henderson, R., & Clark, K. (1990). Architectural innovation. *Administrative Science Quarterly, 35*, 9–30.

Hotelling, H. (1929). Stability in competition. *Economic Journal, 39*, 41–57.

Kirman, A. (1989). The intrinsic limits of modern economic theory: The emperor has no clothes. *Economic Journal, 99*(127), 126–39. Royal Economic Society.

Klepper, S. (2001). *The evolution of the U.S. automobile industry and Detroit as its capital.* Paper for DRUID's Industry Evolution Conference, January 2002.

Klepper, S. (1996). Entry, exit, growth, and innovation over the product life cycle. *American Economic Review, 86*(3), 562–83.

Lancaster, K. J. (1990). The economics of product variety: A survey. *Marketing Science, 9*(3), 189–206.

Lancaster, K. J. (1966). A new approach to consumer theory. *Journal of Political Economy, 4*, 132–157.

Lancaster, K. J. (1971). *Consumer demand: A new approach.* New York: Columbia University Press.

Lipietz, A. (1987). *Mirage and miracles: The crisis of global fordism.* London: Verso Press.

Leffingel, R. (1991). *The American farm tractor: A history of the classic tractor.* Osceola: Motorbook International.

Malerba, F., Nelson, R., Orsenigo, L., & Winter, S. (2003). *Demand, innovation and the dynamics of market structure: The role of experimental users and diverse preferences,* WP CESPRI n. 135, Università Luigi Bocconi, Gennaio.

Marris, R. (1964). *The economics of managerial capitalism.* New York: The Free Press.

Marx, K. (1967). *Capital*. New York: International Publisher. first published 1867.

Mas-Colell, A., Whinston, M. D., & Green, J. R. (1995). *Microeconomic Theory*. Oxford: Oxford University Press.

McKenzie, L. (1959). On the existence of a general equilibrium for a competitive market. *Econometrica, 27*, 54–71.

Morris, W. (Ed.). (1934). *Artist, writer, socialist* (2 Vols. A supplement to the collected works). Oxford: Blackwell.

Murphy, T. (1999), The modularity question. *Ward's Auto World, 35*(4), 53.

Nelson, R. R. (1994). Economic growth via the co-evolution of technology and institutions. In L. Leydesdorff & P. Van den Besselaar (Eds.), *Evolutionary economics and chaos theory: New directions in technology studies*. London: Pinter.

Pasinetti, L. L. (1981). *Structural change and economic growth: A theoretical essay on the dynamics of the wealth of nations*. Cambridge: Cambridge University Press.

Pasinetti, L.L. (1993). *Structural economy dynamics: A theory of the economic consequences of human learning*. Cambridge: Cambridge University Press.

Perloff, J. M., & Salop, S. C. (1985). Equilibrium with product differentiation. *Review of Economic Studies, 52*, 107–120.

Pine, B. J. (1993). *Mass customization :The new frontier in business competition*. Boston: Harvard Business School Press.

Piore, J. M. & Sabel, C. F. (1984). *The second industrial divide: Possibilities for prosperity*. New York: Basic Books, Inc.

Polo, M. (1993). *Teoria dell'oligopolio*. Bologna: Società editrice il Mulino.

Porter, M. (1990). *The competitive advantage of nations*. New York: The Free Press.

Rae, J. B. (1965). *The American automobile: A brief history*. Chicago: The University of Chicago Press.

Rae, J. B. (1984). *The American automobile industry*. Boston: Twayne.

Randall, T., & Ulrich, K. (2001). Product variety, supply chain structure, and firm performance: Analysis of the U.S. bicycle industry. *Management Science, 47*(12), 1588–1604.

Robinson, J. V. (1933). *The economics of imperfect competition*. London: MacMillan.

Sako, M., & Warburton, M. (1999). *MIT international motor vehicle programme: Modularization and outsourcing project—Preliminary report of European research team.* Paper prepared for the IMVP Annual Forum, MIT.

Salerno, M. S., & Diaz, A. V. C. (2002). Product design modularity, modular production, modular organization: The evolution of modular concepts, Les Actes du Gerpisa, 33, Universite' d'Evry.

Saviotti, P., & Pyka, A. (2003). *Economic development, variety and employment.* Working papers, 35, Vienna University of economics and B.A. Research group: Growth and employment in Europe: Sustainability and competitiveness.

Saviotti, P. P. (1996). *Technological evolution, variety and the economy*. Cheltenham: Edward Elgar.

Saviotti, P. P., & Metcalfe, J. S. (1984). A theoretical approach to the construction of technological output indicators. *Research Policy, 13*, 41–151.

SEMA (2002). 2002 *Automotive specialty equipment industry update*. Last access August 20, 2004 from http://content.sema.org/images/pdf/31308.pdf.

Shy, O. (2001). *The economics of network industries*. Cambridge: Cambridge University Press.

Simon H. (1962). The architecture of complexity. In *Proceedings of the American philosophical society* (Vol. 106, pp. 467–482). Cambridge: MIT Press. (Reprinted in *The sciences of the artificial* (2nd ed.) 1981).

Solow, R. M. (1967). The new industrial state or son of affluence. *The Public Interest, 9*, 100–8.

Sorenson, O. (2000). Letting the market work for you: An evolutionary perspective on product strategy. *Strategic Management Journal, 21*, 577–592.

Sraffa, P. (1926). The law of returns under competitive conditions. *The Economic Journal, 36*, 535–550.

Stalk, G., & Hout, T. (1990). *Competing against time*. New York: Free Press.

Stern, C. W., & Stalk, G. (Eds.). (1998). *Perspectives on strategy*. The Boston consulting group. New York: Wiley

Starr, M. K. (1965). Modular production—A new concept. *Harvard Business Review, 43*(November–December), 131–142.

Suarez, F. F., & Utterback, J. M. (1995). Dominant designs and the survival of firms. *Strategic Management Journal, 16*(6), 415–430.

Swaminathan, A. (2001). Resource partitioning and the evolution of specialist organization: The role of location and identity in the U.S. wine industry. *Academy of Management Journal, 44*, 1169–1185.

Swann, P. (2001). The demand for distinction and the evolution of the prestige car. *Journal of Evolutionary Economics, 11*(1), 59–75.

Teubal, M. (1979). On user needs and need determination: Aspects of the theory of technological innovation. In M. Baker (Ed.), *Industrial innovation: Technology, policy, diffusion*. London: MacMillan.

Theil, H. (1972). *Statistical decomposition analysis*. Amsterdam: North-Holland.

Ulrich, K. (1995). The role of product architecture in the manufacturing firm. *Research Policy, 24*, 419–440.

Von Hippel, E. (1986). Lead users: A source of novel product concepts. *Management Science, 32*, 791–805.

Von Hippel, E. (1988). *The source of innovation*. New York: Oxford University Press.

Von Tunzelmann, N., & Wang, Q. (2003). An evolutionary view of dynamic capabilities. *Economie Appliquuee, 56*(3), 33–64.

Weitzman, M. L. (1992). On diversity. *Quarterly Journal of Economics, 107*, 363–406.

Wibbelink, R. I., & Heng, M. S. H. (2001). Evolution of organizational structure and strategy of the automobile industry, *Series Research Memoranda*, Faculteit der Economische Wetenschappen en Econometric.

Wik, R. M. (1972). *Henry ford and grass-roots America*. Ann Arbor: The university of Michigan press.

Williamson, Oliver E. (1985). *The economic institutions of capitalism*. New York: The Free Press.

Witt, U. (2001). Special issue: Learning to consume—A theory of wants and the growth of demand. *Journal of Evolutionary Economics, 11*(1), 23–36.

Womack, J., & Roos, D. (1990). *The machine that changed the world*. New York: Harper Perennial.

## Websites

Advance Auto Parts. (2004). Retrieved August 29, 2004, from http://www.advanceautoparts.com/

Alliance of Automobile Manufacturer. (2004). Retrieved August 28, 2004, from http://www.autoalliance.org

National Automobile Dealer Association (Nada). (2004). Retrieved August 29, 2004, from http://www.nada.org/

Centre for Decease Control and Prevention (CDC). (2004). Retrieved August 29, 2004, from http://www.cdc.gov

General Motor Corporation. (2004). Retrieved August 29, 2004, from http://www.gm.com

Specialty Component Market Association (SEMA). (2004). Retrieved August 29, 2004, from http://www.sema.org

MIX
Papier aus verantwortungsvollen Quellen
Paper from responsible sources
FSC® C105338

If you have any concerns about our products,
you can contact us on
**ProductSafety@springernature.com**

In case Publisher is established outside the EU,
the EU authorized representative is:
**Springer Nature Customer Service Center GmbH**
**Europaplatz 3, 69115 Heidelberg, Germany**

Printed by Libri Plureos GmbH
in Hamburg, Germany